T0350351

SCIENCE UNDER ATTACK

SCIENCE UNDER ATTACK
THE AGE OF UNREASON

RALPH B. ALEXANDER

Algora Publishing
New York

Library of Congress Cataloging-in-Publication Data —

Names: Alexander, Ralph B., author.
Title: Science under attack: the age of unreason / Ralph B. Alexander.
Description: New York: Algora Publishing, [2018] | Includes bibliographical
    references and index.
Identifiers: LCCN 2018035199| ISBN 9781628943634 (soft cover: alk. paper) |
    ISBN 9781628943641 (hard cover: alk. paper)
Subjects: LCSH: Science—Social aspects. | Science—Psychological aspects.
Classification: LCC Q175.5 .A435 2018 | DDC 306.4/5—dc23 LC record available at
https://lccn.loc.gov/2018035199

Printed in the United States

To my parents, Elsie and Solly,
who encouraged me to
challenge conventional wisdom
through scholarship

Table of Contents

Over 200 years have passed since the Age of Reason, or the Age of Enlightenment as it's sometimes called — an era that also marked the peak of the Scientific Revolution which began several centuries earlier. But today, science is characterized more by *un*reason and *un*enlightenment than by the rational discourse of the Age of Reason. Although we pay lip service to science, the emphasis on empirical evidence and logic that distinguish true science from pseudoscience or other imitations is absent from much of the debate over topics such as climate change, vaccination and genetically modified organisms (GMOs).

Apart from the misinterpretation, misrepresentation and even dismissal of evidence, science is currently suffering from a crisis in reproducibility. The situation is so bad that up to 90% of published findings in some areas of biomedical science can't be replicated. Even in psychology, the rate of irreproducibility hovers around 60%. And to make matters worse, falsification and outright fabrication of scientific data is on the rise.

What has brought science to this sorry state? This book examines the forces behind the assault on modern science. To paint a picture of the current state of the discipline, I've chosen six separate examples to illustrate the varied though related forms of attack: continental drift, evolution, dietary fat, climate change, vaccination and GMOs. Each example is a vignette encapsulating the history of the particular topic, together with an analysis of the threats to science, and abuses of science, embodied in the illustration.

Readers may be surprised by my assertion that science is on the side of those who accept the theories of evolution and continental drift, as well as the safety of vaccination and GMO foods but is *not* on the side of those who adhere to the conventional wisdom on climate change — nor those who believe that a low-fat diet can ward off death from coronary heart disease.

In the case of climate change, the actual empirical evidence for a substantial human contribution to global warming is flimsy. The whole case for catastrophic consequences of man-made climate change hinges not on scientific observations but on artificial computer models that have a dismal track record of predicting the future. Like creationism, the notion of dangerous anthropogenic global warming is based on faith, not science. In fact, like many others, I'm struck by the similarities between creationists and advocates of climate change orthodoxy in both their beliefs and their tactics.

Popular belief links climate change skeptics and creationists together as anti-science "deniers." But, while the percentages of global warming skeptics and anti-evolutionists are almost the same, the two groups don't consist of the same individuals. Similarly, the percentage of the public rejecting a major human role in global warming is about the same as the percentage rejecting the consensus among scientists that GMOs are safe to eat, but again both groups of skeptics are quite different.

Consensus in science is necessary for the elevation of a hypothesis to a widely accepted theory. However, the consensus can be mistaken, as shown by several of the book's illustrative examples in which individual and minority views have overcome an apparent scientific consensus to the contrary. Despite the broad initial consensus on the dietary fat hypothesis linking saturated fat to heart disease, a growing body of conflicting evidence has overturned the consensus just 50 years later. And the purported 97% consensus among climate scientists that global warming is largely a consequence of human activity is not only highly exaggerated, but is also on very shaky ground because the empirical evidence for the claim is so weak. In fact, of the six illustrations in the book, the only two in which the original consensus has endured so far are vaccination and GMO foods.

Another striking aspect of consensus on questions such as evolution, climate change, vaccination and GMOs is the gap in opinion between scientists and the general public. With the possible exception of climate change, a far larger percentage of the population in general question the prevailing wisdom on scientific issues than do scientists. The dichotomy highlights one of the reasons for the current attacks on science — its perceived relevance (or irrelevance). The book concludes with a short discussion of the underlying causes of the assault on modern science, and whether it can survive.

I would like to thank physicist and environmental advocate John Droz, Jr., whose 2013 slide presentation "Science Under Assault" was the inspiration for this book. And I owe a big debt of gratitude to my long-ago PhD advisors, Nick Stone at the University of Oxford's Clarendon Laboratory and Geoff Dearnaley (1930–2009) at the UK Harwell Laboratory, for teaching me how to seek the truth among complexity. Maryann Karinch was a great source of encouragement and helpful, honest advice, both in seeking a publisher and in building a platform for the book. At Algora Publishing my editors, Martin DeMers and Andrea Secara, made valuable comments on the book title and manuscript and offered several suggestions for improvement. I thank the whole editorial team at Algora for their painstaking efforts in knocking my technically corrupted manuscript into shape.

Above all, I thank my wife Claudia for her encouragement ("Which chapter are you writing now?") and loving tolerance of the restrictions my writing has imposed on our time together.

Chapter 1. Background

Science today is under attack like never before. While the discipline overall is alive and well and still the source of stunning technological advances, science is crumbling at the edges, because it is being abused, corrupted and even ignored altogether. The magnificent edifice of the scientific method, which dates back more than two millennia, is under threat — just as much as the mighty Roman Empire was when barbarians arrived at its gates.

In the simplest terms, the scientific method is a combination of observation and reason. We'll delve into details later, but the essence of the method is the gathering of empirical evidence together with the use of logic to make sense of the evidence. The term "science" itself comes from the Latin word *scientia*, which means knowledge; it is restricted to the natural or material world as opposed to the supernatural or divine. But what we call science is really abstract knowledge involving theoretical concepts rather than just a collection of facts. The related term "technology" is used to describe practical knowledge, which doesn't necessarily require any understanding of the underlying science.

The assault on science is advancing on many fronts simultaneously. In areas such as climate change and dietary fat, both issues that are topical and very much in the public sphere, the challenge to science comes primarily from political and economic forces. In other areas of public concern, for example vaccination and genetically modified foods, the main threat is simply human fear. And in evolution, it is the centuries-old conflict between science and religion that is gnawing away at our concept of science.

To understand what's at stake, in this chapter we'll first review the history of the scientific enterprise.

Where Science Came From

Historians of science generally trace its origins to ancient Mesopotamia and Egypt, where the development of writing and counting systems enabled permanent record keeping for the first time in human history. Writing not only replaced the previously shaky oral tradition for passing on information and accumulated wisdom, but the new record-keeping capability also made it possible for knowledge gained from observation to be organized. This organization of knowledge in turn improved technologies such as calendar-making. More importantly, it paved the way for the emergence of the sciences of astronomy, mathematics — sometimes called the queen of the sciences — and medicine.

But the science in these and other early civilizations, while it often involved detailed observations, lacked any degree of analysis or abstraction, the distinctive features that characterize science as a discipline. It's the ancient Greeks who first pursued natural knowledge as an end in itself and are regarded as laying the foundations of both abstract science and the scientific method. Technology, on the other hand, has always been an integral part of human existence.

Most of us are familiar with the great names of Greek scholarship. These include the philosophers Plato and Socrates, scientists Aristotle and Archimedes, mathematicians Euclid and Pythagoras, and astronomer Ptolemy among others. Yet it is the lesser-known Thales (c. 624–546 BCE), a philosopher and businessman, who is usually credited with being the "first scientist." Thales achieved this renown because of his ability to reason and make accurate observations in explaining the physical world, in contrast to then more traditional explanations rooted in mythology. His powers of scientific observation were reputedly so astute that he was once able to corner the market in olive presses after correctly anticipating the timing of a forthcoming harvest.

The so-called Hellenic era in ancient Greece, spanning the years from about 600 BCE to the death of Alexander the Great in 323 BCE, is considered by many Greek historians to be the Golden Age of Greek arts and science — a period when Greek scientific thought flourished, and when it most closely resembled our modern scientific method. It was a time when the Greeks cultivated science for its own sake, for the sake of acquiring knowledge rather than for practical applications. Rational debate about nature, perhaps inspired by the penchant of the ancient Greeks for political debate, became common. The very term "natural philosophy" or the philosophy of nature dates from this period.

Unlike today, it was also a time when public funding for science and scientists didn't exist. During the Golden Age, Greek natural philosophers were either independently wealthy or were able to earn a living as private teachers, doctors or engineers. We'll return to the funding issue later in this chapter.

The flowering of Greek science that started with Thales embraced many thinkers over more than two centuries. Among them was Democritus (c. 460–370 BCE), who was one of the creators of an early atomic theory which bears an uncanny resemblance to Dalton's modern atomic theory in the 19th century. However, the two natural philosophers in Greece who stand head and shoulders above all the others, in terms of their contributions to science and their ability to synthesize intellectual concepts, are Plato and Aristotle.

Plato (428–347 BCE), a pupil of Socrates, founded his famous Academy in Athens, where its citizen members met to discuss philosophical problems and escape the hot Greek sun in a shady olive grove, clad in their colorful tunics. More philosopher than scientist, Plato downplayed the importance of experimental observation. Nevertheless, he made a major and lasting contribution to the scientific method by advancing the study of logic, in addition to proposing theories of matter and planetary motion. Although the record of his Academy lectures has been lost, his thirty or so surviving written dialogues illustrate the power of Plato's intellect in many areas including political thought, theoretical science and mathematics.

But the scientific giant of the Hellenic period is Aristotle (384–322 BCE), whose work was a wellspring of ideas that lasted until the modern Scientific Revolution almost 2,000 years later. In contrast to Plato and even more than Thales, Aristotle emphasized a common sense approach and recognized the necessity of experimental observation — the crucial element of the scientific method — based, in his view, solely on the five senses of sight, touch, hearing, smell and taste. His observations about motion led him to theories that prevailed until the time of Newton in the 17th century. He also carried out empirical investigations in what today we call biology, dissected animals, and made significant advances in other areas including matter, astronomy, psychology and the logic of deduction. Tutor to Alexander, Aristotle had his own Athenian academy, the Lyceum. Both his Lyceum lectures and numerous technical treatises have survived, most of his works having been translated into several languages.

Long before the discovery of the chemical elements and their organization into the periodic table, both Plato and Aristotle had proposed that there were just five fundamental elements: the terrestrial elements earth, water, air and fire, and a celestial aether. Their evidence was found in the world around them and in observable phenomena such as steam from boiling water and lightning in the skies. A quaint notion today, it nonetheless illustrates how science began as a quest for an abstract understanding of nature. While technology, in the form of practical skills necessary for survival, goes back to humans' earliest toolmaking and embraces metallurgy and crafts such as weaving, it was the ancient Greeks who first contemplated abstract theoretical questions about the natural world.

Another important aspect of the modern scientific method is skepticism, the constant questioning of observations and conventional assertions. Although not a feature of the Platonic or Aristotelian schools during the Golden Age, skepticism formed an important part of the Hippocratic medical tradition in that period. It also became prominent in the Academy some time later, after Plato's death in 347 BCE.

The medicine of Hippocrates (c. 460–375 BCE) emphasized cautious observation, critical thinking, and the need to be ever vigilant in seeking out sources of error[1] — the basic features of the scientific method. Hippocratic medical theory associated four bodily "humors" or fluids with the four terrestrial elements postulated by Plato and Aristotle: blood with air, phlegm with water, yellow bile with fire, and black bile with earth. Good or bad health was seen as a different balance between these humors. Though again a quaint notion to us today, this ancient view actually survived into the 19th century when modern medicine was born.

We might expect that the major scientific advances of the 300-year Hellenic era would have continued after Plato and Aristotle, especially with the expansion of Greek influence into new kingdoms in Egypt and Mesopotamia. Indeed, for at least the first 100 years after Alexander's death, Greek science retained its vitality and originality. The most preeminent figures during this time were probably the mathematical genius and talented engineer Archimedes (287–212 BCE) and the astronomer Aristarchus (310–230 BCE). It was Aristarchus who put forward the first heliocentric (sun-centered) model of the solar system, similar to that proposed by Copernicus 1,800 years afterward.

This period also saw the founding in the Egyptian capital of the famous Museum at Alexandria, a research institution which employed over 100 scientists and scholars. The Museum, established with state patronage and financial support — a novelty at that time — lasted for 700 years. And it built up a library, which also incorporated Aristotle's, thought to have contained more than half a million papyrus scrolls.

---

[1] Garrison, Fielding H. 1966. *An Introduction to the History of Medicine*. Philadelphia: W.B. Saunders Company, 94.

But the end of the 3rd century BCE marked the beginning of relative stagnation in Greek science, at least according to some science historians. Others date the decline from about 200 CE. Either way, it is well known that the primary emphasis of the Museum at Alexandria wasn't on new science but on consolidation of existing knowledge gained during the Golden Age and shortly afterwards. As Alexandria was the major center for science in both the Greek and later Greco-Roman eras, this tendency to look backward rather than forward would certainly have slowed the pace of scientific advance.

Two notable exceptions were the 2nd-century astronomer Ptolemy (c. 100–170 CE), who conceived a geocentric (earth-centered) model of the solar system which held sway for 1,500 years, and the renowned scientist-physician Galen (c. 129–200 CE). Galen once served as a surgeon to Roman gladiators and was later the Emperor's physician. He conducted his own medical research in several areas and made voluminous contributions to the medical literature of his time. Building on the legacy of Hippocrates and the four humors that regulated health, Galen proposed a physiology of the body in which organs such as the liver and brain exuded "spirits" or essences that controlled bodily functions — a picture that endured for many centuries.

Ptolemy and Galen aside, several reasons have been advanced for the stagnation of Greek science. The predominant view is that because science lacked a distinct role in Greek society, it eventually became regarded merely as an intellectual activity for the idle rich. With the possible exception of the Museum at Alexandria, there was no link between science and its practical applications. Science and technology were thought of as separate pursuits, the more so because of the availability of cheap slave labor that enabled ordinary citizens, rich and poor, to avoid the manual work involved in activities such as mining, metalworking and agriculture, which they despised.[1]

Technology, nevertheless, continued to progress throughout the Hellenic and subsequent eras of ancient Greece, especially during the period following the Roman conquest in the 2nd century BCE. The Roman Empire was remarkable for its many technological achievements, including road building and the development of a durable and waterproof cement which was utilized in the celebrated Roman arch for construction of buildings, bridges and aqueducts. Considerable engineering effort also went into military hardware such as the torsion-spring catapult. It is equally remarkable that the Romans were able to become the technology leaders of the ancient world while paying

---

[1] Farrington, Benjamin. *Greek Science.* 1981. Chester Springs, PA: Dufour Editions, part two, foreword.

almost no attention to science itself,[1] quite the opposite of the modern world. Few Greek writings were ever translated into Latin. During the latter days of the West Roman Empire, not only did science stagnate but the scientific spirit of the Hellenic Golden Age essentially disappeared altogether — and would not revive until the Scientific Revolution in the Middle Ages.

It was once thought that most Greek scientific advances were forgotten in the intervening period, and to some degree this was true in Europe. But it wasn't true in the East, where science was kept alive for several hundred years after the 476 CE fall of Rome, first in the Byzantine (East Roman) Empire and Sassanid Persia, and then in the Islamic world as well as India and China. By 900, Islam had acquired a vast number of Greek science manuscripts from Byzantium and translated them into Arabic. Even today, supposedly more of Aristotle's works and Greek commentaries on them exist in Arabic than in any European language. By 1100, there were ten times more scientists in Islamic countries than in Europe, a level comparable to that in Greece before the Roman conquest.

Indeed, science flourished in the medieval Islamic world. But despite the acquisition and translation of essentially all the major Greek scientific works, including those of Hippocrates and Galen, the primary Islamic emphasis was less on scientific knowledge as an end in itself but more on practical utility.[2] Medieval Islamic scientists made great strides in fields such as astronomy, medicine and optics, especially in connection with vision, and further fleshed out the scientific method established by the ancient Greeks. In medicine, medieval Islam was known for its large network of teaching and research hospitals, and many scientists earned their living as court physicians. Islam is also credited with important progress in mathematics, including the development of modern algebra and the introduction from India of what we now call Arabic numerals.

In India and China during this period, there was less scientific activity than in the Islamic world, but technology marched forward as it did elsewhere. The Chinese in particular made many technological advances which included suspension bridges, paper-making, gunpowder, porcelain china and even the humble wheelbarrow.

---

[1] Notable exceptions were the Roman philosopher Lucretius, whose epic poem *De Rerum Natura* (On the Nature of Things) endorsed the earlier atomic theory of Democritus; and Pliny the Elder, whose extensive documentation of fauna and flora became the principal reference source in natural history until the 16th century.
[2] Ofek, Hillel. 2011. "Why the Arabic World Turned Away from Science," *The New Atlantis*, Winter 2011, 3-23, http://www.thenewatlantis.com/publications/why-the-arabic-world-turned-away-from-science. Accessed 11 July 2018.

Much of the astronomy practiced in China and India, as well as Islam, was directed toward astrology. While today we regard astrology as an irrational pseudoscience, ancient civilizations saw it as knowledge-based, its predictions of future events being derived from observing the apparent motion of stars in the sky. In the context of the time, it therefore constitutes what we label as science, if not natural philosophy. In the same category is alchemy, which was highly developed in medieval Islam as well as China, though often for medicinal applications as much as its false promise of turning ordinary metals into gold or silver.

Yet, just as Greek science ran out of steam after the Golden Age, so too did Eastern scientific activity fade — perhaps during the 12th century in the Islamic world, a little later in China and India. The reasons are varied and still debated among science historians, but there's no doubt that science worldwide entered a period of slumber. It would not awaken for hundreds of years. Although technology continued to make steady progress, including invention of the printing press, the study of science during this interval was restricted to a small number of isolated scholars and university teachers.

The Scientific Revolution

After lying dormant for many centuries, the science that thrived during the time of Plato and Aristotle in the Greek Golden Age gradually came back to life in the 16th and 17th centuries. It was sparked by what is commonly known as the Scientific Revolution. The upheaval of the Scientific Revolution went beyond the reinvigoration of ancient science, however, because it involved the overthrow of Aristotelian and other Greek ideas, in addition to a complete restructuring of the role of science in society.

For reasons that aren't yet fully understood, the Scientific Revolution was confined to Europe. No counterpart occurred anywhere else. Even though the Italian Renaissance in the arts was undoubtedly a factor in spurring the revival of abstract scientific thought in Europe, what is surprising is that western Europe had been an intellectual backwater until at least 1100 — hence the term "Dark Ages." In only a few hundred years, Europeans were able to come from behind and leapfrog the rest of the world in scientific endeavor.

The seeds of the surge in intellectual activity were sown by the founding of multiple European universities. This began with the University of Bologna in 1088 and was soon followed by now famous research universities such as Paris, Oxford and Cambridge. One of the developments that made such a burst of university growth possible was a European agricultural revolution that took place in the Middle Ages, with its accompanying prosperity and widespread urbanization. Another important factor was burgeoning activity in translation of Greek and Islamic science which quickly brought Europe up to speed on past knowledge.

Following the fall of the West Roman Empire in 476, the few Greek scientific works translated into Latin by the Romans had been copied by Catholic priests and scholars in European monasteries — the laborious task no doubt aided by ample supplies of monastic beer or wine — and thereby kept alive for posterity. In 1085, the Christians captured Toledo in Spain from the Muslims and established Toledo as a major translation center. Translation of both Greek and Islamic science into Latin subsequently mushroomed. By 1200, when universities were on the rise, nearly all the works of Aristotle, Hippocrates, Archimedes, Euclid and other ancient Greek luminaries had been translated, along with many Arabic treatises and books. Translation of Greek and Islamic science enabled the new universities to include logic and natural philosophy in their liberal arts curriculum, especially for graduate students in theology, law and medicine, greatly boosting awareness and knowledge of ancient science among society's professionals.

Aristotle's ideas rapidly became dominant and were widely taught. At the same time the Roman Catholic Church, which had established many of the new universities, felt threatened by the sudden popularity of this pagan Greek thought. Aristotelian philosophy frequently clashed with traditional Catholic theology and Church teaching — for example, the Aristotelian notions that the world wasn't created but is eternal, and that the human soul isn't immortal. The Church's response was to attempt to synthesize Christianity with the teaching of Aristotle, a task largely accomplished by theologian Thomas Aquinas (1225–74). Unsurprisingly, it was found that Aristotelian natural philosophy and Catholic theology, which was based on a literal interpretation of the Bible, could not always be integrated successfully. The policy adopted by the Church was that in the event of discord, secular natural science should be subordinated to theology, a solution that only set the stage for future problems.

Thus began a conflict between science and religion, a conflict which persists to this day and which we will revisit in Chapter 3. In medieval times it led to the so-called condemnation of 1277, when the Bishop of Paris condemned 219 "errors" being taught by Aristotelians in the university's arts faculty, together with several books on magic and witchcraft. Backed up by the pope, the bishop threatened anyone who continued to teach the errors with excommunication. Some scholars regard the condemnation of 1277 as the beginning of the Scientific Revolution, normally associated with Copernicus in 1543, because they see the condemnation as effectively endowing intellectuals with the freedom to explore non-Aristotelian approaches to natural philosophy.

Whether this interpretation is valid or not, the event did open the door for medieval scientists to advance new hypotheses about the physical world — so long as they remained hypotheses and didn't threaten the authority of the Church. The concession by theologians gave a big boost to the science that was reemerging in the 13th century. Unfortunately, a succession of crises in the next century, including famine, war and the plague or Black Death, which wiped out as much as a third of Europe's population, completely disrupted the continent for more than 100 years. Eventually, however, both the way of life and scientific scholarship recovered.

The Scientific Revolution, though an important concept in history, wasn't so much a single event as a transition over a short period from the natural philosophy of the ancient Greeks to the science and scientific method of our modern world. Nonetheless, the Scientific Revolution is often epitomized by the Polish churchman and astronomer Nicholas Copernicus (1473–1543). Copernicus published a book[1] just before his death, but written many years before, reviving the sun-centered heliocentric theory of the solar system and universe. The same year saw publication of another significant book symbolic of the scientific transformation taking place: the anatomical drawings of military surgeon Andreas Vesalius, which reflected the increased severity of wounds resulting from gunpowder weapons and other advances in military technology.

A heliocentric theory of the universe had first been proposed by Greek astronomer Aristarchus 1,800 years before, but had been rejected as implausible. This was partly because of the technical problem of stellar parallax.[2] Subsequent astronomers all adopted the earlier earth-centered geocentric theory, originally developed by Plato and Aristotle and later simplified by Ptolemy. Ptolemy's geocentric model endured both because his writings were subsequently translated into Arabic and Latin, and because

---

[1] Copernicus, Nicholas. 1543. *De Revolutionibus Orbium Coelestium* (*On the Revolutions of the Celestial Spheres*), English translation, http://www.webexhibits.org/calendars/year-text-Copernicus.html. Accessed 11 July 2018.

[2] In a heliocentric theory, the earth orbits the sun. Because of the earth's motion, the pattern of stars in the sky should therefore change over a period of six months as the sky is viewed from opposite ends of the orbit. In a geocentric theory, such stellar parallax does not arise because the earth is fixed. No stellar parallax was observed by the ancient Greeks or later astronomers, casting serious doubt on the heliocentric model. But in fact, it wasn't possible to observe stellar parallax at all until the 19th century, when sufficiently accurate optical instruments were first developed.

A heliocentric model can also explain retrograde motion of the planets much more simply than a geocentric model. Retrograde motion refers to the apparent looping motion of a planet in the sky over a period of time.

the medieval Catholic church embraced his view of the universe. But, as we saw above, the Church was also willing to tolerate scientific hypotheses — which included Copernicus' heliocentric theory — simply as hypotheses. Or at least the Church had been willing, until Galileo put his oar into the muddy scientific waters.

The heliocentric model proposed by Copernicus was regarded by astronomers at the time as simpler and mathematically more elegant than Ptolemy's. But it was still marred by the stellar parallax problem, a difficulty only resolved by later, extensive modifications to the model. These were made in a succession of remarkable scientific advances at the hands of Tycho Brahe, Johannes Kepler and Isaac Newton.

Galileo Defies Authority

In 1610, Italian mathematics professor Galileo Galilei (1564–1642, known by his first name) published the booklet *Starry Messenger*[1] reporting utterly unprecedented astronomical observations. He made the astounding observations with a telescope he had developed the year before. While the telescope had actually been invented in Holland in 1608, Galileo soon improved on the basic design and, cleverly recognizing its potential for commerce as well as science, presented an 8-power instrument to the Venetian Senate. The Senate, duly impressed, rewarded Galileo with a salary increase and university tenure.

What was unprecedented were Galileo's first-ever sightings of mountains and valleys on the moon's surface, a total of at least four moons circling the planet Jupiter, and myriads of previously unseen stars in the Milky Way. Initially met with disbelief, his observations not only revolutionized 17th-century astronomy, but more tellingly they presented an unignorable challenge to the conventional scientific wisdom of the day — which was firmly grounded in Aristotelian teaching and Ptolemy's geocentric model of the universe. The astronomical theories of the ancient Greeks all explained the apparent motions of the sun, moon, planets and stars in terms of perfect spheres. But a moon pockmarked with craters and protuberances (Galileo's description) didn't fit this picture at all. Nor did extra moons orbiting neither the earth nor sun. Galileo's observations, as he was quick to point out, were far more easily explained by the sun-centered heliocentric theory of Copernicus postulated 67 years earlier than by the earth-centered Ptolemaic model.

---

[1] Galilei, Galileo. 1610. *Sidereus Nuncius (Starry Messenger)*, English translation, http://homepages.wmich.edu/~mcgrew/Siderius.pdf. Accessed 11 July 2018.

Immediately, publication of Galileo's work created conflict with the Church, which at the time upheld the earth-centered view of the universe presented in the Bible. In 1613, Galileo published another treatise[1] about his telescopic discoveries. This time he reported on sunspots — unquestionably a blemish on the sun's perfection — as well as Saturn, and the phases of Venus. Emboldened by his growing fame, he claimed in the work that his observations on Venus completely repudiated the Ptolemaic system and verified that of Copernicus.

Meanwhile Galileo had given up his university professorship, for which he had little enthusiasm, and taken up a new position as court mathematician for the wealthy Medici family in Florence. Always a contentious personality, Galileo was constantly involved in disputes with other members of the court and with Aristotelian academics. The disputes were over various scientific issues but primarily his telescope observations and promotion of Copernicanism.

In disagreeing with both Aristotelians and the Church by endorsing a sun-centered solar system, Galileo was fighting a widespread consensus. University academics and the educated professionals they taught adhered to Ptolemy's earth-centered theory. Uneducated and illiterate members of the populace believed what they were told by the Church, which was the authority of the time and which stood behind the earth-centered system too, albeit for different reasons. Even the few astronomers who had confirmed what Galileo had seen through his telescope weren't all united in supporting the Copernican picture that eventually gained universal acceptance. But a resolute Galileo insisted that "In questions of science, the authority of a thousand is not worth the humble reasoning of a single individual." Later in the book, we'll encounter other, modern-day examples of individual and minority views overcoming an apparent scientific consensus to the contrary.

One of Galileo's many court disputes led to his writing a lengthy letter in 1615 to the Grand Duchess Christina, mother of his patron. The letter fiercely defended his views and castigated theologians and others who refused to accept Copernican heliocentrism in place of the traditional Aristotelian–Ptolemaic theory supported by the Church.[2] Although written to Christina, the letter was clearly intended for Galileo's critics, to whom the Grand Duchess forwarded copies. A provocative document, it charged

---

[1] _____, 1613. *Istoria e Dimostrazioni Intorno Alle Macchie Solari e Loro Accidente (History and Demonstrations Concerning Sunspots and Their Properties)*, http://portalegalileo.museogalileo. it/egjr.asp?c=36304. Accessed 11 July 2018.

[2] _____, 1615. Letter to Madame Christina of Lorraine, Grand Duchess of Tuscany, English translation, http://inters.org/galilei-madame-christina-Lorraine. Accessed 11 July 2018.

his opponents with being hypocritical and "showing a greater fondness for their own opinions than for truth."

Early in the letter, Galileo took the Church to task for invoking the Bible to settle scientific questions, accusing theologians of abandoning "reason and the evidence of our senses" for biblical passages. This particular accusation is noteworthy, as it reveals in the very same letter not only that Galileo had the insight to have absorbed the essence of the scientific method — based on logic and observation — from the ancient Greeks, but was also discerning enough to recognize that other Greek concepts such as the geocentric theory weren't necessarily valid. Galileo argued in his letter that science and the Bible must agree on questions about nature, since both come from God. In the event of an apparent contradiction, however, he proposed that science should supersede theology — the exact opposite of the official church doctrine devised by Aquinas several centuries before.

Galileo was now playing with fire. Not surprisingly, his earlier publications had attracted the attention of the Roman Inquisition, the organ of the Catholic Church set up to combat heresy, and the Inquisition's Cardinal Bellarmine had commissioned several reports on Galileo's activities. In 1616, Galileo was summoned before Cardinal Bellarmine in Rome. He was told that Copernican theory had been officially declared heretical by the Inquisition and that he could no longer "hold or defend" it, either orally or in writing. Galileo, wisely for now, decided to acquiesce to this restraint, later receiving a formal certificate of confirmation from Bellarmine.

The combative Galileo had not given up, though he turned his attention to other scientific topics in the meantime. On Copernicanism, he simply bided his time until an intellectual cardinal with a strong interest in the arts and sciences, Maffeo Barberini, was elected Pope Urban VIII in 1623. The following year, Galileo spent six weeks in Rome and visited the new pope several times in the Vatican gardens for extended discussions. Although the Pope refused to withdraw the 1616 restraining order as Galileo had hoped, he did approve a book project in which Galileo could revisit heliocentric theory, subject to two conditions. The two conditions were that the book present the Aristotelian–Ptolemaic and Copernican systems as alternative hypotheses and that it emphasize God's ability to move heavenly bodies in unimaginable ways, making it impossible for humans to detect the true cause of events. The Pope himself chose the title, *Dialogue Concerning the Two Chief World Systems*. Galileo was optimistic about being able to convince the Church and the world that he had been right all along but, as it turns out, his optimism was badly misplaced.

When the book finally appeared in print in 1632,[1] Pope Urban VIII was outraged. Galileo had created a tome that advocated strongly for Copernicanism, in defiance of the Pope's requirement for impartiality and despite, at the insistence of the official censor, toning down the preface and conclusion to make the book appear more neutral. Worse still, Galileo had expressed the Aristotelian view of an earth-centered universe through the character of a simpleton (Simplicio), the book being modeled after Plato's dialogues as a conversation between three characters. The other two characters represent Galileo himself, presenting the Copernican view, and an intelligent, undecided man in the street. In the *Dialogue*, written in Italian (*Starry Messenger* had been in academic Latin) to reach the largest possible popular audience, a defiant Galilean character constantly refutes Aristotle and makes Simplicio look stubborn and ignorant. Adding insult to injury, at the end of the book Galileo puts into the mouth of Simplicio an argument about God's omnipotence that the Pope himself had made in 1624 to refute Galileo's theory of the tides. Naturally, the Pope felt that Simplicio was an attempt to ridicule him.

The reaction from the Church was swift and far-reaching. Printing was stopped, an effort was made to buy back copies of the book, which had quickly sold out, and Galileo was ordered to appear before the Inquisition in Rome despite being 68 and in failing health.

Galileo's trial began on April 12, 1633, without his being told of the specific charges. When interrogated about his meeting with the now deceased Cardinal Bellarmine in 1616, Galileo produced the written certificate he had from Bellarmine forbidding him to hold or defend the theory of Copernicus. But the prohibition still allowed him to present the theory as a hypothesis, which is what Galileo claimed, somewhat deceptively, that he had done in his *Dialogue*. His letter from Bellarmine, however, must have rocked the chief prosecutor. The prosecutor had in his possession an apparently different document from Bellarmine, dated three months earlier, enjoining Galileo not to "hold, *teach*, or defend in any way" the sun-centered Copernican view of the solar system. Scholars think this document may be a forgery produced by Galileo's enemies, as it isn't signed by Bellarmine, by any witnesses, or by Galileo himself. The word "teach" was mysteriously missing from Galileo's signed version. Nonetheless, the discrepancy was enough at the time for the prosecutor to formulate what we now call a plea-bargain agreement.[2]

---

[1] Galileo, 1632. *Dialogo Sopra i Due Massimi Sistemi del Mondo* (*Dialogue Concerning the Two Chief World Systems*), English translation, http://www.famous-trials.com/galileotrial/1010-dialogue. Accessed 11 July 2018.

[2] Blackwell, Richard J. 2006. *Behind the Scenes at Galileo's Trial*. Notre Dame: University of Notre Dame Press, 5-6, 13-16.

As his part of the plea-bargain for a reduced sentence, Galileo confessed to "vain ambition, pure ignorance, and inadvertence" and to overstating his case in the book. But shamefully, having obtained this concession, the Inquisition then decided to proceed against him anyway with formal charges of suspicion of heresy — although three of the Inquisition's ten cardinals didn't sign the final sentence. After being convicted and threatened with torture, Galileo recanted his belief in Copernicanism and was sentenced to house arrest for the rest of his life. He was fortunate not to have been charged with full heresy, the punishment for which was burning at the stake. But legend has it that an ever defiant Galileo had the last word, touching the earth and uttering "And yet it moves" as he returned home for his confinement.

In his remaining years Galileo went back to earlier work, making important contributions to the physics of falling objects and projectiles, a precursor to Newton's famous laws of motion formulated in the 17th century. Galileo's *Dialogue* was banned by the Catholic Church, a ban that remained in place until 1835. Only in 1992 did Pope John Paul II issue a statement admitting that the Church had erred in condemning Galileo.[1] Such is the lot of a pioneer, a brilliant scientist ahead of his time, who has been compared to Michelangelo in creative ability and is regarded as the father of modern science. The geocentric theory was finally demolished by Newton in 1687.

The Age of Reason

The Age of Reason, sometimes called the Age of Enlightenment, describes an era in the 17th and 18th centuries distinguished by an emphasis on reason and logical analysis. It marked a pronounced reaction to earlier thinking based on either religious faith or superstition. A period when traditional authority was questioned on a large scale, it embraced not only an upsurge of new and often radical ideas but also social upheavals such as the American and French Revolutions, and the growth of democracy. It was also the period when the Scientific Revolution, which began with the heliocentric theory of Copernicus in 1543, reached its peak and the foundations of the scientific method, originally developed by the ancient Greeks, were reestablished.

Following the dramatic trial of Galileo, science waned in Italy and the Scientific Revolution shifted northwest to France, Holland and England. Enormous scientific progress was made in the next hundred years, mostly in mathematics, astronomy, mechanics and optics, but stretching also to medicine and biology. One of the greatest scientific geniuses of all time lived

[1] Robinson, Daniel N. 2006. "Science and Faith: The Warrants for Belief." In Daniel N. Robinson, Gladys Sweeney and Richard Gill, eds., *Human Nature in Its Wholeness: A Roman Catholic Perspective*, Washington: The Catholic University of America Press, 169.

during the Age of Reason: Sir Isaac Newton (1642–1727), who gave us his celebrated laws of motion and gravity,[1] carried out groundbreaking work in optics, and was the co-inventor of the calculus.

The Scientific Revolution and the Age of Reason saw a decline of interest in the occult, something that even famous scientists such as Newton dabbled in. Newton had a secret alchemical laboratory and believed that alchemical recipes were encoded in ancient Greek myths. Late in life, he predicted that the apocalypse would occur in the year 2060, based on his interpretation of the Bible and a belief that he was one of only a select few who could read hidden biblical codes. As the occult was concerned with supposedly hidden causes of natural phenomena, in the spiritual or even supernatural realm, Newton may have sought through the occult to understand the underlying cause of gravity[2] — something he had acknowledged in his magnum opus on the law of gravity that he could not explain. And it was probably appealing to Newton that occult thinking had no time for Aristotelian science, which allowed observation only of qualities that could be seen, felt, heard, smelled or tasted. Newton's own laws had dealt the final deathblow to the Aristotelian–Ptolemaic geocentric theory of the solar system.

The occult "sciences" of the day included astrology, alchemy, demonology and magic, among others. Both astrology and alchemy had been an essential part of the scientific enterprise from the dawn of civilization. But in Europe, the prominence of astrology had already faded before the Age of Reason, possibly because of opposition from medically trained doctors to its increasing use in medicine; its importance declined even more with the rise of rationality, though astrology by no means disappeared.[3] Magic, on the other hand, in its traditional sense of claiming to alter the course of nature, was doomed by the rational discourse and critical thinking of the Age of Reason; it had previously found its principal outlet in *medical* treatments. Medicine was beginning to mature by this time, following advances such as Harvey's discovery of blood circulation in 1628 and the development of smallpox inoculation in the 18th century. Alchemy, like astrology, had thrived since Babylonian times and was still a robust area of study when the Age of Reason began, at least in England. Yet it gradually gave way to modern experimental chemistry, which had its origins in the early 19th century.

At the same time that scientific activity in the 17th and 18th centuries was beginning to surpass the heights reached in the days of Plato and Aristotle,

---

[1] Formally, the law of gravity is known as the universal law of gravitation.

[2] NOVA video. 2005. "Newton's Dark Secrets," 15 November, http://www.pbs.org/wgbh/nova/physics/newton-dark-secrets.html. Accessed 11 July 2018.

[3] Monod, Paul Kléber. 2013. *Solomon's Secret Arts: The Occult in the Age of Enlightenment*. New Haven: Yale University Press, chap. 2.

the Age of Reason saw a major departure from the Greek view that science and its practical applications are unconnected. During the Golden Age, science was concerned with purely abstract theoretical ideas and, as we saw earlier, Greeks cultivated science solely for the sake of acquiring knowledge. Those who indulged in scientific discussion felt themselves above what they considered the menial work involved in technology. The novel concept that emerged during the Scientific Revolution was that science can in fact be closely linked to technology and can also play a useful role in society — a concept that is still with us today. During the Age of Reason, the medieval view that science should be subservient to theology was finally discarded also.

Oddly enough, the idea of the social utility of science sprang in part from the widespread interest in the occult at the onset of the Age of Reason. Alchemy was then a popular pursuit: Philip II of Spain (1527–98) and Charles II of England (1630–85) both had substantial alchemical laboratories. And Renaissance magic included a fascination with then extraordinary devices such as magnets or mirrors that could produce unbelievable effects.

More significantly, however, English philosopher and statesman Sir Francis Bacon (1561–1626) introduced the concept of science not only as *knowledge* of nature but also as *power* over nature. Bacon saw science as a vehicle to exploit nature's resources for the benefit of humankind. But contrary to some later interpretations, Bacon did not advocate unbridled exploitation of natural resources for profit:

> Lastly, I would address one general admonition to all; that they consider what are the true ends of knowledge, and that they seek it not either for pleasure of the mind, or for contention, or for superiority to others, or for profit, or fame, or power, or any of these inferior things; but for the benefit and use of life; and that they perfect and govern it in charity.[1]

A prodigious intellect, Bacon is also well known for his contributions to the development of experimental science. In emphasizing precise, experimental observation and inductive reasoning, Bacon helped lay the groundwork for the modern scientific method, though his critics point out that a Baconian approach underestimates the importance of imagination and hypothesis that underpin the scientific method of today. Nevertheless, it was during the Age of Reason that experimental science began to flourish.

---

[1] Bacon, Francis. 1623. Preface, *Instauratio Magna (The Great Restoration)*, English translation, https://en.wikisource.org/wiki/Instauratio_Magna/Preface_(Spedding). Accessed 11 July 2018.

The Age of Reason was also the time when the first learned scientific societies and national academies of science — which are primarily professional organizations as opposed to the ancient philosophical academies of Plato and Aristotle — were formed. First and foremost of the new organizations was England's Royal Society. The Society received its royal charter in 1662 and was founded specifically to promote understanding of the natural world through observation and scientific experiment. Shortly afterward, the first periodical journals appeared, as a means of disseminating scientific knowledge and research. And out of scientific publishing evolved the time-honored peer review process for evaluating papers on new scientific findings.

The Scientific Method

The modern scientific method, conceived over two thousand years ago by the Hellenic-era Greeks, then almost forgotten and ultimately rejuvenated in the Scientific Revolution, wasn't refined into its present-day form until well into the 19th century. However, even earlier scientists such as Galileo and Newton had followed the basic principles of the method. By the end of the 19th century when the scientific method had matured, entire new fields of scientific inquiry had opened up, such as electricity, heat, chemistry, biology and the theory of evolution, and the Industrial Revolution was transforming technology and society. It was this explosion of knowledge that gave birth to today's narrow specialization in scientific subjects.

As we've seen, the scientific method combines empirical evidence with reasoning. The basic steps of the method can be summarized as: (1) Observation or data gathering; (2) Formulation of a hypothesis, which is a reasoned explanation of the observations, a sort of sophisticated guess; (3) Testing of the hypothesis by experiment; (4) Verification, rejection or modification of the hypothesis, and retesting if necessary; (5) Independent replication of the experimental results; and (6) Elevation or addition of the hypothesis to a theory, which explains a multitude of confirmed observations, verified hypotheses, and laws — a law describing what happens in specific circumstances. Confusingly, the term "theory" is often applied loosely to a hypothesis alone.

While this prescription may seem sterile and hidebound, in reality the scientific method is much more complicated and less strictly defined. For example, the summary says nothing about how to do the testing. Among the many questions that arise are what type of experiment or experiments should be carried out; which variables should be controlled; how should the experiments be conducted — on a laboratory benchtop or in a full-scale field test, by one person or a team, by which particular method; is an indirect method needed; is cost an issue, and does the cost have to be justified to a funding agency; are ethics involved, as in certain biological or genetic experiments; and so on. Sometimes testing may have to wait for years or even decades, until a suitable test method or better equipment becomes available. If it turns out that the proposed hypothesis needs to be modified to fit the test results, there will in general be more than one way of doing that. And even the interpretation of the experimental data itself may be unclear.

Because the raw data obtained by observation are fundamental to the scientific method, it is important that the data be handled according to certain unwritten rules. These rules include examining *all* the evidence, eliminating any bias in the measurements, and using multiple sources of data where possible to minimize the influence of any personal quirks of the investigators. Otherwise, inferences drawn from the data can't be regarded as reliable, solid science.

The gold standard for testing involving human participation is the randomized double-blind test, most commonly used in clinical trials of both medical treatments and pharmaceutical products.[1] This test is designed to eliminate subjective, often unconscious bias on the part of either the researchers or participants, by withholding information from both that could affect the outcome of the test. In a clinical trial, also known as a randomized controlled trial, participants are assigned at random to two groups, one of which receives the actual treatment or drug, with the other receiving a harmless placebo. Neither the researchers nor the participants are told which group the participants are part of until the very end. The double-blind test assures impartiality and prevents any manipulation of the results.

A hallmark of the modern scientific method is the replication step, which in theory at least makes science inherently self-correcting. If the original data can't be reproduced by independent investigators in another laboratory, the data are clearly suspect and the hypothesis immediately founders. In fact, a number of recent cases of scientific fraud have been uncovered largely because the original data were found to be irreproducible. In other cases, lack of repeatability reflects carelessness by the original investigator who, driven by pressure to publish results as quickly as possible, overlooked a miscalibrated instrument, a contaminated reagent or a faulty statistical analysis. Replication is central not only to the scientific method but also to the advancement of science itself.

---

[1] The same method is often used in psychology and in market research surveys.

But even though following the scientific method may not be simple in practice, its quintessential elements are still observation, hypothesis and reason. The approach has perhaps been described most succinctly by talented U.S. physicist Richard Feynman (1918–88), who said in a lecture on the scientific method:

> Now I'm going to discuss how we would look for a new law. In general, we look for a new law by the following process. First, we guess it [audience laughter], no, don't laugh, that's the truth. Then we compute the consequences of the guess, to see what, if this is right, if this law we guess is right, to see what it would imply and then we compare the computation results to nature or we say compare to experiment or experience, compare it directly with observations to see if it works.
>
> If it disagrees with experiment, it's WRONG. In that simple statement is the key to science. It doesn't make any difference how beautiful your guess is, it doesn't matter how smart you are, who made the guess, or what his name is... If it disagrees with experiment, it's wrong. That's all there is to it.[1]

Feynman won a Nobel Prize for his work in theoretical quantum mechanics, taught undergraduate physics with flair, and also wrote a famous report highly critical of NASA management when the space shuttle Challenger broke up shortly after lift-off in 1986, killing all seven astronauts aboard.[2] In his lecture, Feynman was talking more about hypothesis testing than the observations themselves and, strictly speaking, was addressing development of a new theory rather than a law. But his statement makes very clear the paramount importance of experimental observation to the scientific method — and this from a theorist. Without observation, it's not true science.

---

[1] Feynman, Richard P. 1964. From a lecture given at Cornell University, https://www.youtube.com/watch?v=b240PGCMwV0. Accessed 11 July 2018.

[2] Feynman, Richard P. 1999. *The Pleasure of Finding Things Out: The Best Short Works of Richard P. Feynman*. Cambridge, MA: Perseus Books, chap. 7.

Richard Feynman was a member of the NASA commission that investigated the disaster. Although a theoretical physicist, he actually conducted his own impromptu experiment to test a hypothesis he had formed about the accident. He hypothesized that it occurred because a rubber O-ring seal on a rocket booster lost its flexibility in the near-freezing temperatures the morning of the launch — a possibility shuttle engineers had been concerned about since the discovery of O-ring erosion in a previous shuttle flight. The rigidity of the O-ring would have allowed burning gases to escape, and in fact Feynman had noticed a small flame near the booster seal in a launch photo. This flame, he proposed, heated the main liquid fuel tank, causing it to rupture and burn. Inspired by a glass of ice water at his hotel dinner table the night before, Feynman dramatically demonstrated to his fellow commissioners how an O-ring could freeze and stiffen, verifying his hypothesis. His simple experiment was not only a graphic example of the scientific method at work, but also a damning indictment of NASA's tragic failure to listen to its own engineers.

In formulating and testing hypotheses, two of the standard reasoning processes employed are induction and deduction. Induction, a "bottom-up" process that reasons from a particular case to a general conclusion, was promoted by Bacon during the Age of Reason. Bacon, however, did not connect induction with our modern notion of a scientific hypothesis. In contrast, "top-down" deduction — which was a central feature of Aristotelian science — reasons from general evidence or a theory to a specific conclusion. Deductive reasoning is often used to test the predictions of a hypothesis, while induction is more useful in developing hypotheses. As we'll see in later chapters, both processes can be abused.

Extensive use of the logical processes of induction and deduction in the scientific method has led philosophers to compare the method to detective work in legal investigations.[1] There are certainly parallels, especially at the discovery and hypothesis forming stages, as good detectives are skilled at making what are often unexpected observations and are creative in coming up with hypotheses about a crime. But the analogy can't really be pushed beyond that point, because a detective generally lacks the opportunities and capabilities available to a scientist to perform experimental testing of his or her informed guesses.

Another similarity between science and the law is the crucial dependence of both on evidence and logic. Without evidence, courtroom lawyers don't have a case. And just as in the courtroom, inference from circumstantial or indirect evidence often plays just as important a role in the scientific method as does direct evidence.

An often overlooked feature of the modern scientific method, in addition to the basic steps outlined above, is the falsifiability criterion. This criterion was introduced in the early 20th century by philosopher Sir Karl Popper (1902–94), who grappled with the question of how to distinguish real science from pseudoscience or nonscience. Popper's falsifiability criterion states that a true scientific theory or law must in principle be capable of being invalidated, of being disproved, by observation or experiment.[2] Twentieth-century scientific genius Albert Einstein (1879–1955) is reported to have said about his theory of relativity:

> No amount of experimentation can ever prove me right; a single experiment can prove me wrong.[3]

---

[1] Copi, Irving M., Carl Cohen and Kenneth McMahon. 2015. *Introduction to Logic*. Boston: Pearson, chap. 13.

[2] Popper, Karl R. 2014. *The Logic of Scientific Discovery*. Eastford, CT: Martino Fine Books.

[3] Wikiquote. 2018. "Talk: Albert Einstein," https://en.wikiquote.org/wiki/Talk:Albert_Einstein. Accessed 11 July 2018.

While Popper's criterion is met by all the scientific theories discussed in this book, we'll see there are hypotheses that aren't testable, so can't be falsified and are therefore unscientific.

Closely related to the falsifiability criterion is scientific skepticism, a questioning of scientific claims which relies heavily on evidence and logic, and therefore is an inherent element of the scientific method — particularly the replication step. Scientific skepticism dates back to the Greek Golden Age and the teachings of Hippocrates, and became prominent once more in the Age of Reason. The crucial importance of skepticism was aptly summarized during the 19th century by famous British biologist Thomas Huxley (1825–95), in a remark that echoes Galileo's comment on authority 250 years earlier:

> The improver of natural knowledge absolutely refuses to acknowledge authority, as such. For him, scepticism is the highest of duties; blind faith the one unpardonable sin.[1]

But in case this discussion has left the impression that empirical observation and reason are all there are to science, I should add that the imagination and creativity involved in groundbreaking scientific advances more often than not come from *irrational* thinking by scientists who are passionate about their work. As Einstein once said, "I didn't arrive at my understanding of the fundamental laws of the universe through my rational mind." In a provocative landmark book on the philosophy of science, Thomas Kuhn (1922–96) explored how revolutions in scientific thought occur and challenged the rationality behind acceptance of new theories and paradigms in science.[2] He insisted, however, that science itself is not irrational.

Lastly, as in any kind of endeavor, the scientific method is highly dependent on the human quality of integrity. In Feynman's words,

> ... scientific integrity, a principle of scientific thought that corresponds to a kind of utter honesty — [is] a kind of leaning over backwards. For example, if you're doing an experiment, you should report everything that you think might make it invalid — not only what you think is right about it: other causes that could possibly explain your results.... If you make ... an elaborate theory, you want to make sure, when explaining what it fits, that those things it fits are not just the things that gave you

---

[1] Huxley, Thomas H. 1866. From a lay sermon "On the Advisableness of Improving Natural Knowledge," delivered at St Martin's Hall, London, 7 January. https://ebooks.adelaide.edu.au/h/huxley/thomas_henry/advise/. Accessed 11 July 2018.

[2] Kuhn, Thomas S. 2012. *The Structure of Scientific Revolutions*. Chicago: University of Chicago Press. Kuhn used the term "paradigm" to mean the theories accepted by a particular scientific community, together with the community's consensus on research methodology and on the appropriate questions to ask.

the idea for the theory; but that the finished theory makes something else come out right, in addition.[1]

Science is no more immune to corruption than any other calling. Unfortunately, the last few years have seen an ever increasing number of scientists bending the rules to their advantage and indulging in plagiarism, falsifying of data, or blatant cheating. As we might expect, corruption of science and the scientific method is particularly prevalent in the public arena, an issue that we'll return to later.

Modern Science And Funding

Probably the most far-reaching transformation in science from antiquity to today has been the way it's funded. In the Greek Golden Age, when Plato and Aristotle held leisurely philosophical and scientific debates with their pupils, their academies received no public financing whatsoever and there were no scientific institutions.

That changed in a big way, however, once the Museum at Alexandria was established. Patronized first by the generous Ptolemaic kings and later by Roman emperors, the Museum became a research institution incorporating scientific luxuries such as well-equipped lecture halls and an observatory, in addition to its vast book collection. It also had a sizable payroll for the times of scientists and scholars. Although the purpose of the Museum's research isn't entirely clear, at least some of it seems to have been directed toward practical applications, a forerunner of modern science funding. But from the time of the Greeks until the modern era, technology had always been funded independently of science, initially by city-state rulers and then by nation states.

By the time of the Scientific Revolution, when science reemerged after its long hiatus during the late Middle Ages, a different form of funding had arisen: the court patronage system in medieval Europe. The Renaissance courts of kings and queens, popes and powerful families all employed a vast variety of courtiers, which included physicians, astronomers and astrologers, alchemists, mathematicians, engineers, artists and musicians — as well as scientists, then known as philosophers. Physicians, astronomers and alchemists had also been prominent in earlier Islamic courts. Among medieval European scientists who were financially subsidized by the court patronage system were Danish astronomer Tycho Brahe (1546–1601), who

---

[1] Feynman, Richard P. 1974. From "Cargo Cult Science," a commencement address at the California Institute of Technology (Caltech), http://www.cs.ucsb.edu/-ravenben/cargocult.html. Accessed 11 July 2018.

paved the way for the discovery of elliptical orbits in the solar system, and Galileo as we've seen.

European court patronage of scientists slowly yielded to a system of institutional support. In this new system, scientific experts were supported by the learned scientific societies which sprang up in the Age of Reason, by newly founded national astronomical observatories, and by other state bureaucracies. Yet science and technology remained largely separate and separately funded activities until the 19th century, well after the Scientific Revolution. It was the upheaval of the Industrial Revolution that finally brought science and technology together.

But despite this intertwining of science and technology in the modern world, it is still the practical applications of science that receive the lion's share of the funding available. In the Middle Ages, it was military technologies that dominated spending, with enormous state outlays for advanced weaponry and sailing ships. Today, governments in the industrial world devote less of their total research funds to defense. Yet a large portion of the nondefense remainder is spent on applied science and technology as it is now called, with much less allocated to basic or pure research. This is the polar opposite of public funding for science in the time of the Museum at Alexandria, when far more was spent in studying abstract theoretical ideas than on technology. Although most research today is in fact funded by industry, industrial research is strongly oriented toward technology also, as we would expect.

Nevertheless, what is important in examining the assault on science isn't so much the nature of scientific funding, but its *sources*. The financing of scientists in the past by rulers or powerful families, both during the lifetime of the Museum at Alexandria and during the Middle Ages in Europe under the court patronage system, came without strings attached. At the Alexandria museum, the Ptolemies and their Roman successors honored the Hellenic tradition by allowing scientists essentially complete intellectual freedom, with no direct constraints on their activities. Glory and prestige from the fame acquired by those they funded were adequate rewards. The same was true of medieval court patrons; the Medicis in Italy were happy to employ the talented but controversial Galileo, even if they were later unable to protect him from the Inquisition.

But it's another story today. The freedom enjoyed by scientists in the past to roam at will intellectually is becoming a luxury, and funding by present-day sources is almost invariably accompanied by conditions and restrictions. In the industrial world this is only to be expected, a typical condition being that a research grant should benefit the company in some specific way by producing a result or an idea of value. Companies are in business after all to make money.

In the public sphere, which involves taxpayer dollars, there's more flexibility but grant recipients are usually required at a minimum to write a publicly available report on their research findings. And therein lies the rub. Because public funding actually comes from the government, be it national or local, and governments are political, the door is open to political chicanery and corruption. If a research report is politically incorrect, even though the science may be sound, the scientist risks incurring governmental wrath and being cut off from future funding. Should he or she be a government employee, they may be sidelined or even fired. Conversely, a politically correct research report may guarantee an income stream, but often at the expense of the science involved. U.S. President Dwight Eisenhower (1890–1969) anticipated this very problem in his 1961 farewell address to the nation:

> In this [technological] revolution, research has become central; it also becomes more formalized, complex, and costly. A steadily increasing share is conducted for, by, or at the direction of, the Federal government.

> Today, the solitary inventor, tinkering in his shop, has been overshadowed by task forces of scientists in laboratories and testing fields. In the same fashion, the free university, historically the fountainhead of free ideas and scientific discovery, has experienced a revolution in the conduct of research. Partly because of the huge costs involved, a government contract becomes virtually a substitute for intellectual curiosity.... The prospect of domination of the nation's scholars by Federal employment, project allocations, and the power of money is ever present — and is gravely to be regarded.[1]

Public funding of scientific research creates a conflict of interest between the integrity of scientists on the one hand and their pay and career advancement on the other, as we'll see further on in the book.

In the next chapters, we'll examine six separate examples of the assault on modern science, each one illustrating different though related forms of attack. These examples, which I have chosen to cover the fields of geology, evolutionary biology, nutrition science, climate science, medicine and agricultural science, are representative but by no means all-encompassing. There are many others.

---

[1] Eisenhower, Dwight D. 1961. From his Farewell Address, 17 January, http://www. americanrhetoric.com/speeches/dwightdeisenhowerfarewell.html. Accessed 11 July 2018.

Chapter 2. Continental Drift: A Threat to the Establishment

In this chapter we'll look at a 20th-century controversy that was more of a scientific skirmish than an outright attack on science, as a prelude to the bigger battles described later in the book. In fact, this particular skirmish was ultimately resolved, and the controversy reveals the scientific method operating as it should, albeit on a drawn-out timescale. At the same time, it's also a modern example of the overturning of conventional scientific wisdom by an individual fighting established authority, just as Galileo did in the 17th century with the geocentric theory. But while the Church in Galileo's time upheld the prevailing paradigm by citing the Bible, the scientific establishment in the present case propped up its position by stooping to misuse and abuse of science.

The topic is continental drift, which is the currently accepted theory of how the world's landmasses came to be located where they are.

The Continental Drift Story

Our story begins with a revolutionary geological theory proposed by a young German meteorologist, Alfred Wegener (1880–1930), in 1912. Wegener hypothesized that Earth's continents had not always been where they are today, but slowly drifted to their present positions after breaking away from a supercontinent about 300 million years ago. In the supercontinent, which stretched from pole to pole, the landmasses were all clustered together. This notion was a radical departure from the consensus among geologists of the time that the continents were rigidly fixed in place and had been since time immemorial — a notion that had existed and been taught for almost a century. An eminent U.S. geologist of that era remarked that: "If we are to believe in Wegener's hypothesis we must forget everything which has been learned in the past 70 years and start all over again."[1] He was unintentionally right, but it took almost another 70 years for the theory to be accepted.

Underlying Wegener's imaginative postulate was observation which, as we've seen, is at the root of the scientific method. Wegener had noticed two years earlier, in studying a world atlas belonging to a colleague, that the coastlines of West Africa and South America fit together like pieces of a jigsaw puzzle. This had been observed by others before him and is often rediscovered today by schoolchildren learning geography. But no one had previously explained the observation satisfactorily. In addition to the close match between Africa and South America, Wegener found a not-quite-so-good fit between Europe and North America.

That a meteorologist should have even been interested in geology, let alone put forward a groundbreaking theory, can be attributed to Wegener's passion for exploration. He had dreamed of exploring the Arctic while growing up. While a university student, he had prepared himself for possible Arctic expeditions by scaling mountains, extensive skiing and skating excursions, and long treks in the snow.

After graduating with a Ph.D. in astronomy in 1904, Wegener decided astronomy wasn't for him after all, partly because it wouldn't offer him any opportunities for physical exertion. So he opted instead for meteorology, in which new technologies such as the telegraph and radio were modernizing forecasting and storm tracking, and he went to work at the Aeronautic Observatory outside Berlin. There, Wegener used kites and tethered balloons to probe the upper atmosphere. Ever the adventurer, he also flew in hot-air balloons, breaking the world endurance record in 1906 with his brother Kurt by staying aloft for 52 hours. The same year, Wegener's kite and balloon experiments in meteorology earned him an invitation to join a Danish expedition to Greenland's unmapped northeast coast, an invitation he was thrilled to accept.

Meteorological research on polar air that Wegener carried out during the two-year expedition subsequently gained him a teaching position in meteorology at the small University of Marburg in Germany in 1908. His stay at Marburg was a highly productive period for Wegener, during which he wrote a meteorological textbook which soon became a standard text throughout the country, and also published numerous scientific papers on

---

[1] Hughes, Patrick. 2001. "On the Shoulders of Giants. Alfred Wegener (1880-1930): Alfred Wegener," http://earthobservatory.nasa.gov/Features/Wegener/wegener.php. Accessed 11 July 2018.

his Greenland research. In addition, he found time to pursue his interest in geology and the continents, sparked by his observation in 1910 of matching Atlantic coastlines.

While browsing in the university library in the autumn of 1911, Wegener came across a paper suggesting that Africa and Brazil had once been connected by a continent-size land bridge, now buried under the Atlantic. Former land bridges connecting far-flung continents, which were part of the geological orthodoxy of the day, were an attempt to explain the well-known observation that fossilized animals and plants from the same historical time period could be found in both South America and Africa. The same was true for fossils found in North America and Europe, and in India and Madagascar. As an example, the remains of a now extinct crocodile-like reptile that swam in freshwater have been discovered only in Brazil and southern Africa; it would have been impossible for the reptile to swim long distances across a saltwater ocean. And relics of a fern-like land plant that disappeared about 250 million years ago[1] have turned up in rocks from all the southern continents, including Antarctica, as well as India and Madagascar; the seeds of this plant were too fragile either to have been blown on the wind or to have survived an ocean trip.[2]

However, the now interdisciplinary Wegener rejected the idea of land bridges because there was no plausible geological explanation for their sinking. Instead, he saw another possibility, in light of what he had just learned about fossilized plants and animals — the possibility that the continents had been joined together at one time. In Wegener's words, "Wherever we have in the past allowed for the sinking of former continental platforms into oceanic deeps, we shall now hold for the splitting and drifting apart of continental bergs."[3]

Wegener first published his views on continental drift in the leading geographical and geological journals in 1912,[4] before setting out on a second Danish expedition to Greenland. His paper included several arguments to bolster his disruptive theory. These included the observation that a certain species of garden snail is found today not only in continental Europe, but also in Ireland, Greenland and Newfoundland. Wegener also refined his

---

[1] The scientific names for the reptile and fern are Mesosaurus and Glossopteris, respectively.

[2] Collins, E. M. 2000. "Continental Drift," slide presentation, https://docs.google.com/presentation/d/1eAl4AzusIOpqbgq2Kt8iPhDpgWm-YOxUUC0_F5ObnPM/embed?hl=en&size=m#slide=id.p4. Accessed 11 July 2018.

[3] Greene, Mott T. 1984. "Alfred Wegener." *Social Research*, 51, 739-761.

[4] Wegener, Alfred. 1912. "Die Entstehung der Kontinente (The Origins of the Continents)." *Petermanns Geographische Mitteilungen*, 58, 185-195, 253-256, 305-309. English translation *Journal of Geodynamics*, 32 (2001), 29-63.

original observation, replacing matching coastlines with continental shelves, which fit together much better than shorelines eroded away by millions of years of wind and waves. After returning from a dangerous but successful Greenland excursion, in which he made meteorological observations and took pioneering photographs of clouds, followed by a short stint in the German army during World War I, he published a longer, book-length paper on the continental drift theory in 1915. Further editions appeared throughout the 1920s.

One of Wegener's arguments for continental drift was similarities between the geologies of Africa and South America, Europe and North America, and other continents now widely separated by oceans. The likenesses include mountain chains and coal beds, in addition to flora and fauna. For example, the multi-ridged Appalachian mountain range in eastern North America mimics the Scottish Highlands as well as mountains in Norway, and the lower, folded Cape Mountains of South Africa imitate the Sierras in Argentina's Buenos Aires province.

Wegener also postulated that mountains were formed either when drifting continents ran into resistance from the ocean floor, causing the continental edge to crumple and fold, or when continents collided with each other, as when India hit Asia and pushed up the Himalayas. Wegener argued that only this crumpling process could have produced the narrow bands of mountains that we see today near the edges of continents. In contrast, the conventional wisdom in the early 20th century held that mountains sprung up on the earth's crust like wrinkles on the skin of a drying apple, as the earth cooled and supposedly shrank from its original molten state — the so-called contraction theory. But if this were so, mountain ranges would be spread out evenly over the earth.

Another argument involved glacial sediments left behind in southern Africa, Argentina, southern Brazil, India and Australia, from the long-ago Permo-Carboniferous ice age. This particular ice age was a very long period of glaciation that peaked around the time that Wegener's proposed supercontinent — named Pangea — existed. Pangea was thought to have subsequently split up into two smaller supercontinents: Gondwanaland, embracing the southern glacial sediment continents, and Laurasia in the north. The identical glacial deposits found in disparate parts of the modern world, including deserts, were compelling evidence for the continental drift hypothesis, even to many Wegener critics.

Two years prior to Wegener's first paper on the radical concept of continental drift, Frank B. Taylor (1860–1938), an American geologist and glaciologist, had published a paper putting forward the same idea.[1] Historians regard the two almost simultaneous proposals as independent, but it is Wegener who is considered to be the father of continental drift theory. While Taylor presented similar though slightly different arguments than Wegener's for mountain creation, Taylor's paper made no reference at all to the fossil and glacial evidence for drift theory cited by the younger German meteorologist. Wegener's papers were far more detailed than what Taylor published and drew evidence from many different scientific fields.[2]

Perhaps not surprisingly, the reaction of geologists all around the world to Wegener's theory was militantly hostile. Not only did his unorthodox theory fly in the face of accepted thinking about Earth's history, but Wegener himself was a maverick — and a meteorologist to boot, who seemed to be attacking the very foundations of geology, in which he had no training. Offbeat ideas in science are rarely accepted without a fight, and continental drift would be no exception; one critic harshly dismissed Wegener's hypothesis as "footloose." His publications on the subject unleashed a torrent of rage and animosity, and his opponents attacked him "fiercely, disdainfully, and personally to the point of defamation."[3] He was unfairly accused of incompetence, dishonest use of data, and plagiarism. Because of this abuse, Wegener, who had moved from Marburg to another teaching position in Hamburg, was unable to find a tenured post at any German university. He finally secured a professorship in meteorology and geophysics at the University of Graz in Austria in 1924.

One of the reasons that Wegener's papers were met with such vehemence was that he was unable to come up with a convincing mechanism by which the continents slide around. In his seminal 1912 paper, Wegener had briefly mentioned two possible types of force that could push continents across the surface of the earth. These were a force caused by Earth's rotation, and a so-called tidal force arising from the gravitational pull of both the sun and moon. Whatever the force actually was, Wegener pictured the continents as plowing through the ocean floor under the influence of the force, like an icebreaker smashing its way through ice sheets. Unfortunately for Wegener, this simple conjecture was shown to be wrong by several of his critics, who calculated that the ocean floor and underlying mantle are much too strong to yield to continents on the move as he proposed. Furthermore, the rotational and tidal forces were far too weak to make continents budge: one of his

---

[1] Frankel, Henry R. 2012. *The Continental Drift Controversy, Vol. I: Wegener and the Early Debate.* Cambridge: Cambridge University Press, 61-74.

[2] Wegener did, however, acknowledge Taylor's contribution in his first paper (see Note above).

[3] Greene, 1984, 754.

detractors estimated that a tidal force strong enough to move continents would stop the earth rotating altogether in less than a year.[1]

Similar criticisms cast doubt on Wegener's proposals for the thrusting up and folding of mountain ranges. His continental drift theory did, however, generate vigorous debate and his ideas were widely discussed during the 1920s. He had a handful of defenders in the geological community, although their presentations at scientific gatherings were disparaged just as much as Wegener's lengthy papers. At a symposium of the American Association of Petroleum Geologists in 1926, which was convened for the sole purpose of discussing the controversial Wegener hypothesis and which Wegener himself did not attend, one of the participants made the sarcastic comment:

> If Wegener or anyone else can throw light on these baffling problems, he is entitled to a hearing. However, certain demands are made of this new and romantic speculation before it is admitted into the respectable circle of geological theories. It must meet the test of established scientific principles, and it must not create more problems than it pretends to solve.[2]

Stung by such remarks and weighed down by trying to champion his revolutionary ideas, Wegener in 1927 responded enthusiastically to the suggestion of yet another Greenland venture. The German expedition, led by Wegener, set out in 1930 after a reconnaissance trip the year before, but unfortunately he never returned. Once again, one of the expedition objectives was to gather meteorological data, this time on the jet stream in order to aid transatlantic aviation. Three weather stations were to be set up: one near the center of the Greenland ice cap, the other two on the west and east coasts. But keeping open the central weather station, 400 kilometers (250 miles) inland, during the bitter Greenland winter proved to be Wegener's undoing.

The expedition fell behind schedule from the beginning, first because of unusually thick pack ice at sea on the west coast and then because of strong winds and heavy snow. The party consisting of Wegener, a colleague, and thirteen Inuks struggled to cope in the -50° Celsius (-60° Fahrenheit) cold, and had to abandon half its supply package for the central station en route. All but one of the Inuks were sent back to the coast at the same point. By the time the remaining trio arrived at their destination after a 40-day journey, Wegener's exhausted colleague had such severe frostbite that he could travel no further. Because they had been forced to dump supplies, there was barely

---

[1] University of California Museum of Paleontology. "Alfred Wegener (1880-1930)," http://www.ucmp.berkeley.edu/history/wegener.html. Accessed 11 July 2018.

[2] Longwell, Chester R. 1928. "Some Physical Tests of the Displacement Hypothesis." In W. van Waterschoot van der Gracht, ed., *Theory of Continental Drift: A Symposium on the Origin and Movement of Land Masses Both Inter-continental and Intra-Continental, as Proposed by Alfred Wegener*, Tulsa: American Association of Petroleum Geologists, 145-157.

enough food and fuel for two, let alone all five at the station. It was decided that Wegener and his Inuk companion would return to the coastal base.

The two set out with a dogsled the next morning, once more in unrelenting cold. Wegener died, most likely of a heart attack, halfway back to the coast. His body, wrapped in sleeping bag covers, was found in a shallow grave, marked with his skis, the following spring. His companion, whose body was never discovered, had evidently buried Wegener, taken his diary and attempted to continue back to the coastal station. Wegener's widow chose to leave his ice-entombed body in Greenland as a fitting monument to the intrepid explorer.

Despite Wegener's death, the debate over his magnum opus on continental drift continued. Yet his ideas began to languish in the 1930s, not because of the absence of Wegener but because of ever increasing parochialism or regionalism among the world's geologists. Those who studied the Permo-Carboniferous ice age, characterized today by its glacial sediments, tended to favor drift theory owing to the abundance of geological evidence — fossilized plants and animals, as well as glacial deposits — that suggest the present-day southern continents were once linked together. On the other hand, geologists in North America and the former Soviet Union saw their local regions as geologically self-contained and had little time for drift theory, clinging to the consensus contraction theory instead.

As a result of this regional factionalism, combined with the lack of a plausible mechanism for continental drift, Wegener's theory was all but ignored for the next thirty or forty years. One bright light in this period of intellectual darkness, however, was a hypothesis advanced by English geologist Arthur Holmes (1890–1965). Holmes proposed a possible mechanism for both continental drift and mountain building, based on the heat generated by radioactive elements in the earth's mantle, which lies just under the outer crust.[1]

It had already been theorized that the breakdown of radioactive elements such as uranium and thorium produces enough warmth to melt the mantle locally; the molten liquid or magma, being less dense than the solid material around it, then rises up to the top of the mantle. At the top, the magma cools off through contact with the cooler crust above, resolidifies, and sinks again, before the cycle starts once more. It was suggested by Holmes that this convection current or conveyor-belt type of process could exert enough pressure from below to shift continents and thrust up mountains in the crust. His idea received little attention at the time, because heating from radioactivity was incompatible with the contraction-theory consensus

---

[1] Frankel, 2012, Vol. I, chap. 5.

of a cooling earth. However, Holmes' proposal actually foreshadowed an imaginative concept proposed in the early 1960s known as seafloor spreading, which shortly afterwards led to a stunning new theory about the ocean crust and continents that finally vindicated Wegener's ideas.

Forty Years On: New Evidence

Forty years after Wegener published his first paper on continental drift, interest in his unorthodox ideas gradually revived in the 1950s when startling new evidence for migrating continents was found. Even though Wegener had bolstered his case for continental drift with observations about animal and plant fossils, glacial deposits, mountain ranges and coal seams, all this evidence was what a detective or lawyer would call circumstantial — indirect evidence from which the drift hypothesis could only be inferred.

The new, more direct evidence consisted of accurate measurements of the earth's magnetic field. As we learn in high school, the earth behaves magnetically like a giant bar magnet aligned with the planet's rotational axis, with north and south magnetic poles close to the corresponding geographic poles. Before the days of GPS devices, we found our way around the world with the help of the compass, the needle of which always points toward the north magnetic pole. As part of an effort after World War II to understand what drives the earth's magnetic field, geologists across the globe set about gauging the field in multiple locales. While earlier measurements had been handicapped by inaccuracy, in 1947 English physicist Patrick Blackett (1897–1974) introduced a much more sensitive instrument which he had designed and which enabled his geological colleagues to get a much more accurate handle on the earth's field than ever before. Blackett had built the instrument to check out his idea, erroneous as it turned out, that the magnetism of planets (including Earth) and stars arose from their rotation.[1]

Both the strength and polarity of the earth's magnetic field vary over time. In fact, every few hundred thousand years on average the polarity completely flips, so that the north magnetic pole now sits at the south geographic pole and vice versa. Geologists are especially intrigued by the history of such magnetic reversals,[2] which don't seem to have any devastating effects on planetary life, because they provide clues to the behavior of the earth's core. Luckily for geology, this history is recorded in many of the world's rocks.

---

[1] Frankel, Henry R. 2012. *The Continental Drift Controversy, Vol. II: Paleomagnetism and Confirmation of Drift.* Cambridge: Cambridge University Press, 1-2.
[2] The study of the past strength and direction of the Earth's magnetic field is known as paleomagnetism. The instrument used to measure the field is called a magnetometer.

It was discovered centuries ago that ordinary bricks become magnetized in the direction of the earth's field at the time they are fired. Much the same phenomenon occurs in igneous or volcanic rocks such as granite and basalt. In these rocks, molded by fire when a volcano erupts and disgorges a stream of lava or magma, tiny iron-rich crystals of magnetic minerals in the molten magma line themselves up with the earth's magnetic field. When the magma finally cools and solidifies into rock, these crystals record a memory of the field strength and direction, like so many tiny compasses.[1] What's amazing is that this record, known as remnant magnetism, is preserved for billions of years.

Sedimentary rocks such as limestone and sandstone, created by the settling of solid residues out of water, also have a magnetic memory. But the magnetism in sedimentary rocks is puny, about 10 times weaker than in volcanic rocks and thus more difficult to measure. This is because sediments often contain less magnetic material than lava flows and because the magnetic crystals aren't locked into place until most of the water evaporates, allowing time for some of them to be jostled out of alignment.

Blackett's supersensitive instrument was just the ticket for the band of geologists and geophysicists in the early 1950s who had set their heart on cataloging variations in the magnetic field across the planet. Although the remnant magnetism in volcanic rocks is strong and easy to measure, volcanic rocks are relatively few and far between since volcanic eruptions occur only sporadically. Sedimentary rocks, on the other hand, are formed constantly through weathering and erosion, and they therefore offer a continuous magnetic record. This continuity in the sedimentary record was important to the geologists studying the earth's magnetic field, who were trying to ascertain exactly when magnetic reversals have occurred in the past. Fortunately for them, the weakness of sedimentary magnetism was no longer a challenge with the advent of Blackett's new device.

Two British research groups spearheading these studies soon made an unexpected observation. Although the direction of the north pole recorded in rocks less than about twenty million years old was close to the present-day direction of the earth's magnetic field, or to its magnetically reversed direction which is flipped by 180 degrees from the present, this wasn't at all true in older rocks. In rocks older than about twenty million years, the memory compasses pointed at a large angle away from either pole direction — sometimes as much as 90 degrees. The unusual observation had two

---

[1] The alignment is locked in only when the magma has cooled below the Curie Point, which is the temperature above which magnets lose their magnetism. If the solidified rock ever becomes heated above the Curie Point once more, the previous magnetic record is lost and a new magnetic memory is recorded when the magma cools again.

possible explanations. First, the north magnetic pole may not always have been in the same place, but may have wandered around until twenty million years ago. Shifting of the magnetic pole, or apparent polar wander as it was labeled, wasn't an unreasonable idea at the time, when the origin of the earth's magnetic field wasn't understood as well as it is today, and a pole that wandered for an extended period was a distinct possibility.

The second potential explanation for the puzzling jumps in the rock record was that the rocks themselves must have moved, from the ancient time when their magnetic memories were recorded to the present day. If the north magnetic pole had always been fixed, in either its normal or flipped position, a change in angle of the rock memory compasses could only mean a change in the latitude at which the rock formed.[1] In other words, the landmasses of which the rocks were part must have shifted by enormous distances and reoriented themselves — Wegener's continental drift theory resurrected. The British rock magnetization results implied that Britain had moved thousands of kilometers north from the time the rocks were created, and had rotated many tens of degrees into the bargain.[2] Similar observations began to be made in other parts of the world including Europe, India, Australia, southern Africa, South America and Antarctica.

Both possible explanations, or a combination of the two, were given serious consideration by the geological community in the 1950s and became the topic of a protracted debate in the scientific literature and at scientific gatherings. But while the drifting continent explanation soon attracted a growing cadre of devotees, especially up-and-coming PhD students in geology, 40 years of entrenched opposition to the continental drift hypothesis would not easily be overcome. Renegade North American and some Australian geophysicists, in particular, favored polar wandering over continental drift to explain the magnetism of ancient rocks, maintaining that the evidence for drift was weak.

Yet by 1960, it had become clear that there was abundant, strong evidence for continental drift in the rock magnetism record. Early measurements from the mid-1950s had shown that the memory compasses of ancient rocks of similar age in Europe and North America not only pointed in a very different direction than that for recent rocks, but the European and North American directions differed from each other. If the continents had stayed in place and the north pole had truly been moving around, these two directions would be close together. Although a combination of polar wander and continental

---

[1] The angle of dip between the direction of the earth's magnetic field and the horizontal, known as the inclination, is directly related to the geographic latitude of the measuring instrument.

[2] Frankel, 2012, Vol. II, 37-41, 62-67, 104-117.

drift could not be completely ruled out, astonishing data from India and Australia left no room for ambiguity. Because the magnetization directions for similarly aged rocks in both countries were so drastically different from those for Europe or North America, the only plausible conclusion was that both India and Australia had drifted very large distances since ancient times. In fact, the data indicated that India had crossed the equator, from the southern hemisphere to its present location in the northern hemisphere.[1]

But despite such impressive evidence, a vocal group of diehards continued to throw obstacles, mostly imagined, in the path for acceptance of continental drift theory. The endless, almost impossible tasks that the die-hard holdouts imposed on drift-theory advocates to verify Wegener's hypothesis have been described as the "Twelve Labors of Hercules."[2] Heavily influenced by powerful and influential voices among the critics of continental drift, the majority of geologists of the day saw the new data on rock magnetism as demonstrating a wandering north pole rather than drifting continents. At a NATO conference in 1963 on ancient climates, only 16 out of 29 presenters whose research papers expressed an opinion on polar wandering or continental drift favored drift theory; most who opposed it were North American.[3]

Nevertheless, the wheel finally began to turn in the 1950s and early 1960s when measurements of the earth's magnetic field in land-based rocks were extended to the seafloor. The seafloor is far from flat, just like landmasses, and dominated by two main features: an extensive chain of volcanic and largely submarine mountains known as mid-ocean ridges, which divide the major oceans roughly in half; and long, narrow and very deep trenches or canyons near the continental edges. Mid-ocean ridges surpass the Himalayas in size, rising up to four and a half kilometers (three miles) from the seabed and spanning a width of more than two kilometers (one mile).

Measuring the magnetism of rocks buried under the sea calls for a different approach from the laboratory techniques used to measure rock samples collected on land. As part of the post-World War II thrust to survey the earth's magnetic field around the world, rugged airborne instruments, based on technology originally developed to reveal military submarines, were utilized initially to map the magnetization of the seafloor. However, flying airplanes was expensive. So geophysicists at the Scripps Institution of Oceanography in the U.S. came up with a much cheaper method, in which one of the submarine-finding instruments was towed by a ship engaged in

---

[1] Ibid, 134-138, 198-208.

[2] Ibid, 426.

[3] Frankel, Henry R. 2012. *The Continental Drift Controversy, Vol. III: Introduction of Seafloor Spreading*. Cambridge: Cambridge University Press, 36-47.

other research activities. This method was subsequently used in extensive surveys of the ocean floor off the Pacific coast of North America.

A survey conducted in 1955 and 1956 revealed bizarre, puzzling results. Although earlier airborne surveys had detected irregularities in the undersea magnetic field associated with atolls and submerged volcanoes, these had been expected. So too had small variations from place to place, because most of the world's oceanic crust is basalt, and basaltic rocks vary slightly in composition and magnetic strength. But what was totally unanticipated in the survey data was a pervasive zebra-like pattern of magnetic stripes, alternating between stripes of normal magnetic polarity — with the tiny memory compasses in the rock pointing toward the north pole — and adjoining stripes of exactly the opposite polarity. Further surveys confirmed the phenomenon over a wide area of the eastern Pacific seafloor and elsewhere, the magnetic striping always running roughly parallel to the mid-ocean ridges.

While no immediate answers could be given to how the mysterious stripe patterns originated, a major step toward an explanation was the concept of seafloor spreading put forward by U.S. geologists Harry Hess (1906–69) and Robert Dietz (1914–95) in 1960 and 1961. The basic idea is that ocean crust is continually regenerated by hot basaltic lava upwelling from depths of tens of kilometers in the earth's mantle, right underneath the crust. Compared with the rest of the ocean floor, the mid-ocean ridges are structurally weak. Because of this weakness, the radioactive heating that turns solid mantle rock into magma exerts pressure from below, ripping open cracks that run lengthwise along the ridge crest. Molten magma then spews out of the cracks, triggering colossal subsea volcanic eruptions and also freezing as new ocean crust right next to the ridge. Geologists think this violent process in fact created the mid-ocean ridges in the first place.

Seafloor spreading is accompanied by the swallowing up of ocean crust in trenches along the ocean rim, far from the central ridges. This effectively recycles the crust and avoids the need to postulate an expanding earth, which had been an alternative explanation for continental drift.[1] As well as throwing light on the genesis of ocean ridges, the seafloor spreading concept explained several observations, including the fact that rocks are relatively youthful near the ridge crests but become progressively older toward trenches.

---

[1] Kious, W. Jacquelyne and Robert I. Tilling. 1999. "This Dynamic Earth: The Story of Plate Tectonics. Developing the Theory," http://pubs.usgs.gov/gip/dynamic/developing.html. Accessed 11 July 2018.

Most importantly, the spreading concept solved the mystery of magnetic striping. It was hypothesized in 1963 by English geophysicists Frederick Vine and Drummond Matthews (1931–97) that the zebra bands of magnetized rock on the seabed, which line up with a ridge crest, are simply a "tape recording" of historical flip-flops in the earth's magnetic field, the magnetic reversals already studied on land and later confirmed in deep-sea sediments. As magma erupted from the ridge and then cooled to form new ocean crust, it would have encoded the direction of the earth's field at that time. This scenario would have continued as further seafloor spreading pushed the new crust outward from the ridge, until the next time the direction of the earth's field switched and was encoded in the newly forming seafloor as a reverse magnetic stripe.

Sixty Years On: Acceptance

Now the stage was finally set for the acceptance of continental drift theory, put forward so long ago by Wegener in 1912. Continental drift could explain both the existence of underwater ridges right in the middle of the oceans and, through seafloor spreading, the spectacular, newly discovered magnetic striping on the ocean floor. Geological weaknesses in Pangea, the ancient supercontinent proposed by Wegener, would have initiated rifting of the supercontinent along what were to become the mid-ocean ridges. Seafloor spreading then separated the individual continents and caused them to move ever further apart until they reached their present-day positions. Because spreading occurs at the same rate on either side of a mid-ocean ridge, the ridges are approximately equidistant from the two rifted continents that border that ocean. This continental drift picture, derived from marine geological observations, was reinforced by all the rock magnetism evidence previously obtained on land.

Like continental drift, however — though with less acrimony — the notion of seafloor spreading was at first met with strong resistance among some in the geological world. One reason was that not until two years after magnetic striping was explained could it be demonstrated that the magnetic stripes migrate outward from any particular ridge at a steady rate, which is the behavior expected in seafloor spreading. Before this discovery, it had appeared that the spreading rate varied from epoch to epoch in the past. The discovery was made possible by a new, more accurate timescale for dating ancient rocks and sediments, based on radioactivity.

But most of the holdouts were won over when it was shown that seafloor spreading could explain not only the origin of subsea magnetic stripes, but also earthquake activity along a certain type of geological fault in the ocean crust, known as a transform fault. Transform faults, which are sometimes thousands of kilometers in length, break up the mid-ocean ridges into zigzag segments.[1] The faults transfer the vertically upward motion of upwelling magma to horizontal seafloor spreading along a fault, and eventually transform horizontal to vertical motion beneath a trench.

Although seafloor spreading clearly implied the existence of continental drift, it did not provide a mechanism. One of the ironies of the whole continental drift saga is that Wegener's hypothesis was eventually accepted in the late 1960s, even though his icebreaker analogy of continents plowing their way through the ocean floor, driven by some unknown mechanism, had been mercilessly shot down by his critics forty years earlier. Seafloor spreading suggests instead that the continents simply ride passively on the underlying crust, which is continually created at ocean ridge crests and dives underneath the continents at trenches.

In any case, the issue of a mechanism for continental drift was sidestepped when the theory of plate tectonics emerged. Plate tectonics describes the motions of 15 to 20 thick slabs of the earth's crust, or hard and rigid overlapping plates, that glide over the mantle in various directions at very slow speeds — typically a few inches per year. The relative movement of these plates explains continental drift as well as a series of observations, including the global distribution of most earthquakes and mountains, the existence of mid-ocean spreading ridges, and other features on the ocean floor such as magnetic striping and transform faults. Mid-ocean ridges, for example, occur where two oceanic plates are pulling apart as hot magma pushes its way up between them. Transform faults on land are exemplified by the famous San Andreas Fault in California, in which oceanic and continental plates are currently sliding past each other.

Plate tectonics says nothing about *why* the plates move. Radioactive heating and the conveyor-belt convection currents in the earth's mantle, first suggested by Holmes in the 1920s, may play a role. But there have been several other suggestions, all of which are still being debated today.[2] Nevertheless, the advent of the theory was enough to silence the few remaining critics of continental drift once and for all; the importance of plate tectonics to geology

---

[1] A ridge-ridge transform fault is a type of strike-slip fault perpendicular to the ridge, which forms where two tectonic plates slide past one another. The ridge offsets created by transform faults are part of the original continuous ridge structure. Because of the offsets, ocean crust on one side of the fracture zone is older and deeper than crust on the other side.

[2] Kious and Tilling, 1999. "This Dynamic Earth: The Story of Plate Tectonics. Some Unanswered Questions," http://pubs.usgs.gov/gip/dynamic/unanswered.html. Accessed 11 July 2018.

has been compared to the importance of the theory of evolution to biology. In stark contrast to the 1963 NATO conference at which 11 out of 29 presenters were opposed to the continental drift theory, the geophysicist who presided at a NASA conference late in 1966 "could not find a single scientist" to defend the belief in fixed continents.[1]

Abuses Of Science

As I've mentioned already, the continental drift story follows the playbook for the scientific method outlined in Chapter 1. Observations were made — of continental geography, of animal and plant fossils found in widely different parts of the globe, of mountain chains, of rocks with a memory of where they were magnetized; a radical hypothesis was formulated to explain the first observations; the hypothesis was initially rejected, then modified after experimental testing that constituted the later observations, and finally accepted as a theory when it was shown that the cornucopia of evidence collected over many years fit the modified hypothesis. That's pretty much how any scientific theory comes into being.

So how was science abused? The abuse actually took several forms, including the vicious personal attacks on Wegener, closed-mindedness among many of his critics, and unreasonable objections to continental drift theory that spanned a period of four decades and delayed its eventual acceptance.

It's perfectly normal in science for scientists to vigorously defend their intellectual turf and to resist new ideas that threaten to overturn the prevailing wisdom, not to mention their pet theories. Indeed, skepticism is an intrinsic part of the scientific method. What was abnormal in the continental drift debate was the barrage of virulent *ad hominem* attacks directed at Wegener while he was alive. German geologists scorned what they called Wegener's "delirious ravings" and other symptoms of "moving crust disease and wandering pole plague." At a meeting of the Royal Geographical Society in England, an audience member thanked the speaker for blasting Wegener's theory apart, and then thanked the absent "Professor Wegener for offering himself for the explosion."[2] Reactionary American geologists painted him as

---

[1] Frankel, Henry R. 2012. *The Continental Drift Controversy, Vol. IV: Evolution into Plate Tectonics.* Cambridge: Cambridge University Press, 412-418.
[2] Conniff, Richard. 2012. "When Continental Drift Was Considered Pseudoscience," *Smithsonian Magazine,* June, http://www.smithsonianmag.com/science-nature/when-continental-drift-was-considered-pseudoscience-90353214/?no-ist. Accessed 11 July 2018.

a fool, a dreamer or a poet, with his maps "not worth the paper on which they are printed."[1]

Such derision, although common in the political sphere, is not science. Neither the science that was debated at length in the academies of Plato and Aristotle, nor the science that reemerged in the Age of Reason, had any place for personal animosity. Even though Newton abused his position as President of the Royal Society in his notorious priority dispute with Gottfried Leibniz over invention of the calculus, he never stooped to the level of personal hostility experienced by Wegener. However, as we will see later in the book, Wegener's shabby scientific treatment in the 20th century may have merely been a prelude to what is almost standard in the 21st.

Although Wegener was modest enough to admit in 1929 that "The Newton of drift theory has not yet appeared,"[2] he was never welcomed by the geological club of his time. With no formal background in geology and more of a modern earth scientist than a geologist — with interests in meteorology, oceanography, paleontology and astronomy, as well as geophysics — Wegener and his migrating continent hypothesis were seen as a huge threat to the geology establishment. Geologists who had risen to power by backing rival theories, in which the earth had always looked as it does today, felt their reputations endangered by this upstart who couldn't even satisfactorily explain why the continents should move at all. He was falsely accused of ignoring facts and using circular arguments. Just how much Wegener's revolutionary theory threatened the establishment in the U.S. is revealed by the reaction of scientific publishers: of seven articles published on continental drift in American geological periodicals in 1924, not one was favorable to his theory.[3]

In battling a scientific consensus that supported an opposing view, Wegener is often compared to Galileo. Galileo had to fight the authority of a church that ranked theology above science, though he himself dutifully followed the principles of the scientific method. Wegener, on the other hand, was pitted against fellow scientists and both sides were ostensibly playing by the same scientific rulebook. Nonetheless, many of his opponents frequently displayed an authoritarian belief in their view of permanently fixed continents.

Wegener's nemesis among authoritarian critics was British geophysicist Sir Harold Jeffreys (1891–1989), who in 1922 was the first to correctly

[1] Newman, Robert P. 1995. "American Intransigence: The Rejection of Continental Drift in the Great Debates of the 1920s." In Henry Krips, J. E. McGuire and Trevor Melia, eds., *Science, Reason, and Rhetoric*, Pittsburgh: University of Pittsburg Press, 181-209.
[2] Greene, 1984, 758.
[3] Newman, 1995, 188.

take issue with Wegener's proposed mechanism for continental drift. But Jeffreys' objections to the theory soon degenerated into a cavalier attitude and unreasoning closed-mindedness toward the revolutionary hypothesis. In the 1929 edition of his treatise *The Earth*, which was one of the standard geophysics textbooks for decades, Jeffreys argued that Wegener's almost perfect fit of South America and West Africa was really a misfit of about 15 degrees. He repeated the criticism in the next edition of his book in 1952. And after an Australian professor of geology had demonstrated in a detailed stereographic analysis in 1955 that the fit was indeed excellent, an obstinate Jeffreys, when asked a few years later if he had examined this fit, said he had never read any of the Australian's research papers and had no intention of doing so.[1] To the end, even after the theory of plate tectonics was accepted in the late 1960s, he remained vehemently opposed to the idea of continental drift.

Jeffreys also rejected the overwhelming evidence for drift accumulated from measurements of rock magnetism during the 1950s, and he even questioned the then newly accepted hypothesis among geologists that the earth's magnetic field is on average aligned with its axis of rotation. Just as in the continental misfit episode, this closed-mindedness on Jeffreys' part reveals that he had probably read few of the papers on rock magnetization. Trained in mathematics with little education in geology as such, Jeffreys was even unaware that geologists only work with rocks whose magnetization isn't changed when beaten with the hammer used to collect samples.

Because of Jeffreys' immense prestige, which stemmed from a wide range of accomplishments in mathematical geophysics, his views on any subject carried weight and therefore influenced his peers. Nevertheless, his refusal to acknowledge the steadily expanding body of data that verified Wegener's continental drift theory was an abuse of both science and power. His main objection to the theory never changed, namely that there was no known mechanism to explain drifting continents or moving plates. But, while spirited debate is an inherent part of the scientific process, observation — the centerpiece of the scientific method — should never be minimized, however surprising or unpalatable the data.

Two of Jeffreys' American disciples, the theoretical geophysicists Gordon MacDonald (1929–2002) and Walter Munk, were just as narrow-minded about the magnetization of ancient rocks as Jeffreys was about continental drift in general. They frequently made fun of the geologists who interpreted the burgeoning number of magnetic memories found in the world's rocks as evidence for moving continents rather than a wandering north pole. In

---

[1] Frankel, 2012, Vol. II, 317-320; Vol. III, 141.

a scathing 1963 review of an anthology on continental drift, MacDonald sarcastically dismissed drift as being an appealing theory only because it was "a favorite topic of pundits condescending to the lay public" and at the same time so complex that it allowed "drifts in styles of thinking." In a further burst of venom, he taunted the theory's supporters for displaying "an appearance of universal genius which mankind has not seen since Elizabethan times."[1]

Here again, science was badly adrift. Not only was MacDonald and Munk's constant denigration of rock magnetism unreasonable, at a time when this emerging area of investigation had gained the respect of the vast majority of field geologists, but their refusal to acknowledge the relentless build-up of magnetization data favoring continental drift was unprofessional and unscientific.

The same unreasonable attitude among American geophysicists to the accumulating evidence for continental drift is enshrined in a widely read 1960 review of rock magnetization data, by Allan Cox (1926–87) and Richard Doell (1923–2008).[2] In a very lengthy and erudite commentary, Cox and Doell went out of their way to build a strong case for polar wandering over continental drift, by cherry picking evidence that favored the shifting north pole explanation — which, thanks to the influence of powerful figures such as Jeffreys, had become the consensus view among geologists and geophysicists at the time. Cox and Doell buttressed their case by emphasizing the weaknesses of the continental drift theory and trivializing its strengths. In a piecemeal approach, they discounted the crucial data from India, which revealed a rock magnetization direction drastically different from that for Europe or North America and which was a key piece of evidence supporting Wegener's drift theory. Similarly convincing data from Australia was downplayed too.

Bias like this in such an important review paper is as much of an abuse of science as the *ad hominem* attacks on Wegener or the narrow-mindedness of his posthumous critics. Not only did Cox and Doell selectively choose their magnetic data to bolster polar wander, but they also ignored the whole history of the continental drift debate. And they neglected to discuss further supporting evidence for drift that was beginning to come from other fields of study such as ancient climates and ancient wind patterns.

Some of the opposition to Wegener's continental drift theory emanated from regionalism, as I discussed earlier in the chapter. The refusal of most North American geologists and geophysicists from the 1920s onwards to

---

[1] Frankel, 2012, Vol. III, 25-31.
[2] Cox, A. and R. R. Doell. 1960. "Review of Paleomagnetism." *Bulletin of the Geological Society of America,* 71, 645-768.

accept the drift concept became institutionalized, and was passed on from generation to generation, from supervisor to student, as well as embedded in textbooks. This intellectual intransigence led to a limited knowledge of geology beyond North America and a lack of interest in the geological and biological similarities between different regions, such as the two sides of the Atlantic. This in turn perpetuated antagonism toward drift theory and generated resistance to new evidence such as the rock magnetism results, and new ideas such as seafloor spreading. An American geology undergraduate in the late 1950s was told that continental drift was impossible and that "serious young scientists shouldn't work on such a stupid idea."[1]

Although regionalism may not be unscientific in itself, the tribal behavior that it represents can get in the way of scientific progress. Another manifestation of tribalism in the continental drift debate was the groupthink that caused the bulk of the geological community to reject the solid evidence for drift provided in the 1950s by the rock magnetism record. Groupthink is not only the enemy of skepticism and a serious threat to science, but also a very powerful facet of human behavior.

The cumulative effect of all these abuses of science was that acceptance of Wegener's theory was delayed by at least 20 and perhaps 40 years, much more than was necessary. While it may be true that his original observations about matching coastlines, fossils and glacial deposits were insufficient building blocks for a complete theory, the discovery of rock magnetism in the early 1950s should have been enough to convince doubters that continental drift was the only reasonable explanation. But antipathy toward the theory, fueled by the vitriolic attacks on Wegener during his lifetime and unreasonable objections by his opponents after his untimely death, was only finally overcome by yet more evidence, this time from the seafloor, and the advent of plate tectonics. The passage of time was possibly a factor too, as expressed by famous German physicist Max Planck (1858–1947):

> A new scientific truth does not triumph by convincing its opponents and making them see the light, but rather because its opponents eventually die, and a new generation grows up that is familiar with it.[2]

---

[1] Frankel, 2012, Vol. IV, 387.
[2] Planck, Max. 1948. *Wissenschaftliche Selbstbiographie (Scientific Autobiography)*. Leipzig: Johann Ambrosius Barth Verlag, 22. English translation by Frank Gaynor, 1949, *Scientific Autobiography: and Other Papers*, New York: Philosophical Library, 33-34.

It has been speculated that had Wegener lived longer, he would have been part of the plate-tectonics revolution, if not the actual instigator.[1] Whatever might have been, however, the rejection of continental drift theory for so long unquestionably dealt a blow to science.

[1] Kious and Tilling, 1999. "This Dynamic Earth: The Story of Plate Tectonics. Alfred Lothar Wegener: Moving Continents," http://pubs.usgs.gov/gip/dynamic/wegener.html. Accessed 11 July 2018.

CHAPTER 3. EVOLUTION AND CREATIONISM: SCIENCE VS RELIGION

This chapter delves into a controversy that has simmered for several hundred years, over the evolution of life on Earth. According to the scientific theory of evolution formulated by English naturalist Charles Darwin (1809–82), the life forms we see around us, including ourselves, evolved largely through the process of natural selection. The radically different, religious belief of creationism, which predates Darwin but survives today, holds that life as we know it was created by a divine designer, perhaps as recently as 6,000 years ago. Darwin's theory of evolution was revolutionary at the time he proposed it and is still not universally accepted, especially in some parts of the U.S., because of widespread adherence to creationism. Yet most established religions don't accept creationism over evolution as an explanation for our origins.

Here the issue isn't abuse or corruption of science, but whether creationism is science at all. Although creationists attempt to argue that their views are scientific, we'll see that the evidence and logic characteristic of true science are lacking.

The Evolution Story

Darwin's theory of evolution represents not only an upheaval in science comparable to that triggered by Copernican heliocentric theory three centuries earlier, but also a triumph for the importance of scientific observation. His book, *On the Origin of Species by Means of Natural Selection, or the Preservation of Favoured Races in the Struggle for Life,*[1] was mostly based on an enormous collection of animals, plants and fossils that Darwin gathered on a long sea voyage in the early 1830s. After observing a large degree of diversity in his collection, Darwin became convinced that biological evolution — or cumulative change over time, involving descent from a common ancestor — must be at play. But while the mechanism of natural selection as the evolutionary driving force occurred to him shortly after the voyage, Darwin held off from publishing his theory until 1859. Like Copernicus before him, Darwin hesitated to go public with a theory that was as unproven and provocative in its day as the sun-centered theory of the solar system.

The concept of biological evolution itself wasn't unknown to Darwin. His own grandfather Erasmus Darwin (1731–1802) was one of the first proponents of the idea of descent with modification, as Charles Darwin described the evolutionary process. Descent implies heredity, and today we know that genes are passed down from one generation of a species to the next in a dynamic process. The elder Darwin's vision of evolution was dynamic also, but Erasmus saw the "inherent activity" of an organism as the evolutionary mechanism, the organism modifying itself in order to adapt to its environment.

Though now a discredited notion, a similar view involving the inheritance of acquired characteristics was promulgated by French naturalist Jean-Baptiste Lamarck (1744–1829). One of Lamarck's best-known examples was the giraffe, whose long neck was thought at the time to be the result of generations of ancestors stretching to reach foliage in high trees, with longer and longer necks then being inherited.

These 18th-century beliefs contrasted with the largely static, creationist worldview that the observed adaptation of complex organisms to their environment was evidence of design by an intelligent creator. In a famous analogy, English clergyman William Paley (1743–1805) argued that coming across a watch on a heath implies the existence of a watchmaker because, unlike a stone found there, an object as complex as the watch with all its cogs and springs must have been assembled by a designer for the specific purpose of telling the time. Structural complexity in nature, of the human eye for example, therefore implies a designer too, with the purpose in this case of providing sight. According to Paley,[2] the designer was God; a similar argument had been made by Aquinas in the 13th century. In such a picture,

[1] Darwin, Charles. 1859. *On the Origin of Species by Means of Natural Selection, or the Preservation of Favoured Races in the Struggle for Life.* London: John Murray, Albemarle Street, http://darwin-online.org.uk/converted/pdf/1861_OriginNY_F382.pdf. Accessed 11 July 2018.
[2] Paley, William. 1802. In Frederick Ferré, ed., *Natural Theology: Selections*, Indianapolis: The Bobbs-Merrill Company, 1963, 3-44.

living creatures including humans were created in their present form and evolution is an unnecessary concept. The young Charles Darwin initially embraced Paley's watchmaker argument and the thesis of creationism, though in time he would turn to the more dynamic, evolutionary ideas of his grandfather Erasmus.

Charles was fascinated by nature from an early age. Sent to study medicine in Edinburgh at age 16, he spent more time collecting and dissecting sea creatures, and learning to stuff birds, than on his classes. This passion for collecting, along with his revulsion at medical procedures, led his father to suggest Charles go into the Church instead, many clergymen being amateur naturalists.

At the University of Cambridge, where he went to study for the clergy, Darwin sought out and established contact with several prominent botanists, one of whom encouraged him to become a full-time naturalist. During his student days, Darwin had a lesson in the hazards of his would-be calling when he discovered three rare beetles after tearing off some old bark from a tree. Holding one in each hand, he tried to put the third into his mouth — only to have it eject a pungent liquid that burnt his tongue, causing him to spit it out and drop one of the others.[1] It was at Cambridge that Darwin began a serious interest in geology and read a book documenting the voyages of a German explorer and naturalist to the tropics of South America. Both these activities would shortly be of value to Darwin as the official naturalist for a British mapping expedition to South America and the Pacific islands, a position he accepted in 1831. His disappointed father was persuaded that this was a better opportunity than becoming the clergyman he was now qualified to be.

Although the unfortunate Darwin was seasick for much of HMS *Beagle's* five-year voyage, he was nevertheless active when on land and energetic enough not only to collect a wide variety of specimens, but also to make a multitude of geological observations and climb the Andes mountains repeatedly. Contrary to many reports of the voyage, geology occupied Darwin's time as much as biology and he had with him on the *Beagle* a copy of a major geological textbook. After experiencing an earthquake off the coast of Chile, he concluded that the whole Andes range had been elevated by accumulated quake action in the distant past. And his observations in the Pacific Ocean of atolls, which are circular coral reefs enclosing a shallow

---

[1] Barlow, Nora. 1958. *The Autobiography of Charles Darwin 1809—1882.* London: Collins, 62, http://darwin-online.org.uk/content/frameset?pageseq=1&itemID=F1497&viewtype=image. Accessed 11 July 2018.

green lagoon in the midst of a deep blue sea, led Darwin to a geological explanation of atoll formation that still stands today.[1]

But it was Darwin's observations of the diversity in wild life and plants, from unusual fungi to elephant-sized fossil bones to a new species of South American ostrich that were the basis of *Origin of Species*. Because the journal he kept of the *Beagle* voyage wasn't published until several years after he returned to England, by which time he had already conceived his theory of evolution, we can't be sure that the journal account accurately reflects Darwin's thinking at the time it was recorded. However, one entry in the published journal does provide an insight into his development of the theory.

The journal entry, prompted by his discovery of geologically recent llama-like fossils on the Patagonian coast, includes the statement: "Certainly, no fact in the long history of the world is so startling as the wide and repeated exterminations of its inhabitants."[2] The same entry points out the close connection that Darwin found between extinct species in South America and the animals currently living there. This recognition, perhaps well after his actual observation, of descent with modification exhibited by creatures in the present-day world was a major step toward Darwin's epochal theory.

Now that Darwin had assembled such a trove of empirical evidence for biological evolution, he needed to come up with a mechanism to drive the evolutionary process. Darwin found his mechanism in the writing of T. R. Malthus (1766–1834), an English priest and political economist whose 1798 *Essay on the Principle of Population* argued that the human population would outpace the food supply were it not limited by famine, disease and war. Malthus maintained that competition for available resources results in a perpetual struggle for existence that keeps the population down.[3] Darwin's creative leap was to extend this idea from humans to all species, being aware that, as in human society, nature produces more offspring than can survive, and having observed on his *Beagle* voyage that variation in a species results in some offspring having a slightly greater chance of survival. These offspring have a better chance of reproducing and passing the survival trait on to the

---

[1] Bowler, Peter J. 1996. *Charles Darwin, The Man and His Influence.* Cambridge: Cambridge University Press, chap. 4.

[2] Darwin, Charles. 1845. *Journal of Researches into the Natural History and Geology of the Countries Visited During the Voyage of H.M.S. Beagle Round the World, Under the Command of Capt. Fitz Roy, R.N.* London: John Murray, Albemarle Street, 172-176, http://darwin-online.org.uk/content/frameset?itemID=F14&viewtype=image&pageseq=1. Accessed 11 July 2018.

[3] Malthusian theory is no longer valid in today's world, due to mechanization and vast improvements in agricultural productivity that Malthus did not foresee. However, the theory was applicable in the 18th-century, before industrialization, when the global population was much smaller.

next generation than those who lack the trait. This is the essence of natural selection, as expressed by Darwin himself:

> In October 1838, that is, fifteen months after I had begun my systematic inquiry, I happened to read for amusement Malthus on *Population*, and being well prepared to appreciate the struggle for existence which everywhere goes on from long-continued observation of the habits of animals and plants, it at once struck me that under these circumstances favourable variations would tend to be preserved, and unfavourable ones to be destroyed. The results of this would be the formation of a new species. Here, then I had at last got a theory by which to work.[1]

Natural selection is often described as "survival of the fittest." Although this description, coined by one of Darwin's contemporaries, was also adopted by Darwin, the true meaning of natural selection is differential survival combined with reproduction, as stated above. It isn't necessarily the biggest and strongest of a species that survive, but rather the individuals that can reproduce more often in the face of an environmental threat, such as a suddenly hotter or wetter climate, a new predator or a mutated virus. Fewer organisms with "less good" genes making them vulnerable to the threat survive than organisms with "good" genes that enable them to survive longer and produce more offspring. This results in a new generation with relatively more good genes, and one therefore slightly better equipped to ward off the threat. As the incremental natural selection process continues over a long period of time, which may be thousands or even millions of years, a new species will eventually emerge. The characteristics of the new species will be recognizably different from, though still related to, the original species.

A common misconception about natural selection, relevant to our discussion of creationism later in the chapter, is that it is an entirely random process. This is not so. Genetic variation within a species, which distinguishes individuals from one another and arises from mutation or other sources, is indeed random. However, the selection part isn't random but rather a cumulative process, in which each evolutionary step that selects the variation best suited to reproduction builds on the previous step.[2] The procedure amounts to a nonrandom sorting that, in Darwin's words above, preserves favorable variations from one generation to the next. What can be said about natural selection is that, while not random, it is a mindless process, one without any particular purpose. In terms of Paley's watchmaker

---

[1] Barlow, 1958, 120.
[2] Dawkins, Richard. 1996. *The Blind Watchmaker: Why the Evidence of Evolution Reveals a Universe Without Design.* New York: W. W. Norton & Company, 43-50.

analogy, as described so eloquently by British evolutionary biologist Richard Dawkins, natural selection is the *blind* watchmaker.[1]

Although Darwin formulated his hypothesis about natural selection in 1838, just two years after he left the *Beagle*, it wasn't until 21 years later that he went public with his findings on evolution. One reason was that he wanted to publish his geological observations and theories. But a far more important reason for Darwin's hesitation to publish his evolution theory was the radical impact he knew it would have on a Victorian society that was still heavily inclined toward Paley's creationist views. He had realized that his multitude of biological observations could only be explained by an evolutionary process, not by divine creation. During this time Darwin became ill, to the point where he could work for only a few hours a day. Some historians attribute his illness to a parasitic infection contracted on the *Beagle* voyage, but it is considered more likely to have stemmed from intense anxiety over the conflict of his nascent theory with public opinion, and even with his own wife's religious beliefs.[2]

In 1858, Darwin began to panic on discovering he was about to be scooped by another naturalist, Alfred Russell Wallace (1823–1913). Wallace, who was largely self-educated, had his own substantial collection of specimens and had begun to publish his views on evolution. In response, Darwin in 1856 had started to write what would become *Origin of Species*, approximately two thirds of which was finished by the time he received the draft of a research paper from Wallace, outlining a completely independent description of natural selection. Fearing that he might not receive credit for the theory of evolution, Darwin leveraged his high-level contacts in the scientific world to arrange for the simultaneous publication of Wallace's paper and a short version of his own, much longer treatise. This allowed both Darwin and Wallace to be recognized for the idea of natural selection, although the theory advanced by Wallace was far more restricted and less comprehensive than that of his colleague. Darwin's full tour de force, which occupied more than 500 pages, came out a year later in 1859.

Nevertheless, Darwin's theory of evolution wasn't widely accepted in his lifetime. Just as he had anticipated, his proposed mechanism of natural selection, even though initially endorsed by Wallace, ran afoul of religious thinking of the day. The church had warmed somewhat to evolution itself and was no longer wedded to the concept of creationism. Yet there was a strong yearning to retain the presumption of a divine watchmaker, who at

---

[1] Dawkins, Richard. 1996. The Blind Watchmaker: Why the Evidence of Evolution Reveals a Universe Without Design. New York: W. W. Norton & Company, 5.
[2] Bowler, 1996, 68-75.

least guided evolution if not the very process of creation, rather than abandon the designer notion for the unpredictable natural selection of Darwinian theory. And Darwin's representation of evolution driven by natural selection as a "tree of life," in which branches decay and fall off or grow into new species, was at odds with the popular view of a simple ladder of progress that ascended toward modern humanity.

Scientific objections were raised against his theory too. At the time Darwin published *Origin of Species* genetics was undiscovered, so nothing was known scientifically about inheritance, and little evidence existed that the earth was more than a few hundred thousand years old. Darwin's forays into geology had convinced him that many geological formations date back several hundred million years — the same span of time necessary for life as we know it to have evolved on Earth through the step-by-step natural selection process.

To placate his critics, Darwin watered down his theory in the next two decades. While he never abandoned his central premise of natural selection, he was willing to concede to his opponents that natural selection might not be the exclusive mechanism behind species change. So he retreated into Lamarckism, which postulates the inheritance of acquired characteristics as the evolutionary driving force, supposedly exhibited by giraffes as we saw earlier. The older Lamarckian view was an alternative to natural selection, allowing those who disagreed with Darwin to retain their belief in an underlying divine purpose for evolution.

As difficult as it was for Victorians to swallow the concept of uncontrollable natural selection, it was the implications of the theory of evolution for humans that provoked the strongest reaction and exposed Darwin to ridicule in the media. At a famous meeting in Oxford of the British Association for the Advancement of Science, shortly after *Origin of Species* appeared, Bishop Samuel Wilberforce insolently asked Darwin's supporter Thomas Huxley if he was descended from an ape on his grandfather's or grandmother's side. Huxley's cleverly biting retort, which created uproar, was that he would rather be descended from an ape than from a *man* who misused his great gifts and religious beliefs to obscure the truth.[1]

---

[1] Huxley, Leonard. 1900. *Life and Letters of Thomas Henry Huxley, Vol. I*. New York: D. Appleton and Company, 192-204, https://archive.org/details/lifeandletterst13huxlgoog. Accessed 11 July 2018.
   Some scholars maintain that the reported retort by Thomas Huxley has been embellished over the years. However, while no actual transcript of the meeting proceedings exists, this biography of Thomas Huxley by his son Leonard includes a lengthy account of several eyewitness reports of the verbal clash between his father and the Bishop — most of which support the account described here. Thomas himself, in an 1891 letter to Darwin's son Francis, states that on hearing the Bishop's "insolent question," he "made up [his] mind to let him have it" (page 202).

Although *Origin of Species* had only hinted at the evolutionary connection between humans and apes, the raging debate that followed its publication prompted Darwin to publish *The Descent of Man*[1] in 1871, in order to clarify that his theory applied to both humans and the animal world. However, in this book he went further than just insisting that evolution had sculpted humanity's physical features, in postulating that it had also shaped mental and social characteristics such as intellect, emotions and behavior. This latter claim gave rise to what is known as social Darwinism, in which natural selection operates in the social as well as the biological sphere. The concept of social Darwinism was fashionable in the late 19th century but later became discredited, as the rise of the social sciences fostered deeper understanding of social and cultural phenomena.

One of the reasons that Darwin's theory, especially his natural selection hypothesis, wasn't well received during his lifetime was that he had no explanation for how the natural selection mechanism actually works. It was commonly assumed at the time *Origin of Species* was published that inheritance incorporates a blending process, in which parental characteristics are blended in the offspring. But blending inheritance would quickly dilute the improved trait through reproduction with unchanged individuals, and would soon eliminate it altogether — nullifying natural selection and preventing the formation of new species.

As we now know from the pioneering experiments of Austrian monk and botanist Gregor Mendel (1822–84), heredity isn't a blending process but transmits traits in discrete units as genes. Since genes are never blended out, beneficial traits are preserved and eventually spread through the whole breeding pool. Mendel's experiments on plant breeding, first described in a paper only six years after the publication of *Origin of Species* in 1859, remained unknown even by the time of Darwin's death in 1882. Mendelian genetics supplies the missing key to how natural selection works, but even after Mendel's paper was rediscovered in 1900, Darwin's theory of evolution languished as the new science of genetics temporarily veered off in a different direction. Lamarckism, first postulated in the 18th century and later revived by Darwin himself, remained popular also. But the tide gradually began to turn as additional evidence for evolution accumulated.

Modern Evidence for Evolution

While the empirical evidence for evolution and natural selection was largely limited during Darwin's lifetime to an incomplete fossil record, this

---

[1] Darwin, Charles. 1871. *The Descent of Man, and Selection in Relation to Sex.* London: John Murray, Albemarle Street, http://darwin-online.org.uk/EditorialIntroductions/Freeman_TheDescentofMan.html. Accessed 11 July 2018.

shortcoming was overwhelmingly rectified in the 20th century with the emergence of completely new fields of study. Of these, the two advances that had the greatest impact on Darwin's theory were the discovery of radioactivity, which led to the radiometric dating process, and the elucidation of the structure of DNA, which led to the field of molecular biology.

Radiometric dating methods rely on the radioactive decay of certain chemical elements such as uranium, carbon or potassium, for which the decay rates are accurately known.[1] Initial acceptance of the theory of evolution was hampered by the prevailing belief that the earth simply wasn't old enough for the incremental natural selection process to have significantly influenced the genetic traits of animals and plants. But in the early 1900s, radiometric dating dramatically boosted estimates of Earth's age from hundreds of thousands of years first to millions and then to billions of years. This was plenty of time for evolution to have taken place.

The revolutionary field of molecular biology enables scientists to sequence the DNA of living species — the coded genetic instructions that allow an organism to form, grow and reproduce. The more recently that the common ancestor of two species lived, the more alike are the genes in their DNA sequences and the more closely related the species are; the more ancient the ancestor, the more unlike is the DNA of the two species and the more distantly related they are. Providing some of the strongest evidence for evolution is the use of an apparent evolutionary relationship, inferred from DNA sequencing and sometimes reinforced by a comparison of the species' physical features, to predict the existence of common ancestors in the *fossil* record. A recent example is the discovery in the 1990s of dinosaur fossils in China that helped confirm the evolution of birds from ancestral reptiles.

Both radiometric dating and DNA sequencing are valuable methods in the field of biogeography, or the study of how different species are distributed across the globe. Biogeography came into its own with the growing awareness of Wegener's theory of continental drift, discussed in the previous chapter. As we've seen, Earth's continents were in very different positions hundreds of millions of years ago, with Africa abutting South America and both India and Australia cheek by jowl with Antarctica. The subsequent splitting of the continents has led to similar fossils, as well as living animals and plants, being found today in geographically disconnected regions.

---

[1] The radiometric dating method measures the ratio of parent (the isotope that decays) to daughter (the decay product) amounts in the sample whose age is to be determined. The age depends on both this ratio and the decay rate or isotopic half-life. Uranium, which is often used to date ancient rocks with ages of billions of years, is also one of the elements responsible for radioactive heating of the earth's mantle, as discussed briefly in Chapter 2.

The enormous diversity of living creatures and plants is another major plank in the observational evidence for evolution. Long before modern tools became available, Darwin himself had remarked that "neither the similarity nor dissimilarity of the inhabitants of various regions can be accounted for by their climatal and physical conditions."[1] One illustration of this is marsupials — pouched mammals like the kangaroo that give birth to very undeveloped young — which are found almost exclusively on the island continent Australia. Placentals — mammals like ourselves with placentas that enable their young to be born at a more developed stage — are found everywhere else except Australia. Despite their anatomical differences, the two species of mammal are remarkably similar in other respects: the marsupial sugar glider, for example, glides from tree to tree just like the placental flying squirrel. Only evolution from a common ancestor, combined with continental drift, can explain this dichotomy. Because Australia was geographically isolated for millions of years, marsupials evolved independently from placentals and, through natural selection, adapted to a different environment. The same is true for animals inhabiting oceanic islands, such as the famous finches that Darwin discovered on the Galápagos Islands during his *Beagle* voyage: the ecologically diverse group of birds evolved in an empty habitat without competition or predators.

Further evidence for evolution comes from the field of embryology. It was already known in Darwin's time that all vertebrates, including humans, surprisingly begin development looking like a fish embryo. As the embryo grows, different species diverge; the structures that become the gills in fish and sharks turn into the heads and necks of mammals such as us. Darwin, who considered embryology, and not fossils, his best evidence for evolution, argued that the sequence of embryonic development in any species mimics the evolutionary sequence of its ancestors. Molecular biology corroborates this view and reveals that as one species evolves into another, the descendant inherits all the genes that formed ancestral embryonic structures. During subsequent evolution of the new species, new genetic instructions are added to the instructions for development of its ancestors, so that a descendant's embryo no longer displays all the features of the ancestor's embryo.[2]

Yet another piece of evidence for evolution is what Darwin called "rudimentary or atrophied organs," or organs that are relics of past body structures but no longer serve any function. His examples from the animal kingdom included "the stump of a tail in tailless breeds," vestiges of hind limbs and hip bones in snakes that are no longer necessary, and unused

---

[1] Darwin, 1859, 302.
[2] Coyne, Jerry A. 2010. *Why Evolution Is True.* London: Penguin Books, chap. 3.

insect wings that are much too small for the insects to fly.[1] Human examples include our tiny tailbones, tonsils and appendix. Such obsolete features or atavisms are readily explained by evolution, as inherited from ancestors who once had functioning versions of the organs. Since Darwin's time, this picture has been amply confirmed by molecular biology. The DNA sequences of essentially all species are found to include "dead" genes or pseudogenes, genes that were useful in the past but have since been silenced, and so normally don't function anymore. Occasionally, however, the embryonic development process goes astray and these dormant genes are reawakened, producing vestigial structures.

Darwinism was finally accepted as the dominant paradigm in evolution during the 1940s, more than 80 years after Darwin announced his theory. That's just a little longer than it took for Wegener's theory of continental drift to gain acceptance in the same century in the face of opposition from the scientific establishment, as discussed in Chapter 2. Opposition to Darwin's theory came primarily from another establishment, that of the Church or, more precisely, 19th-century religious doctrine. As we saw in Chapter 1, Galileo had to battle the established authority of the Church to promote the heliocentric theory of Copernicus some 250 years earlier. However, two centuries make a big difference: the Church in the Victorian era could not punish Darwin for his views as the Inquisition had punished Galileo, but merely spoke out against him. By the mid-20th century, most of the world's major religions had integrated evolution into their theology, with the notable exception of religious fundamentalists in the U.S.

The Creationism Story

Creationism has its roots in biblical literalism, which first gained a significant foothold in 16th-century Europe when the Bible was translated from Latin into multiple contemporary languages and so became more widely read. At that time, belief in the stories of creation and the great flood in the Bible's book of Genesis were just as common as the belief that the sun revolved around the earth. According to Genesis 1, the world and all its living creatures were created by God all of a sudden, in just six days. By the time of the Age of Reason, however, the rise of science was causing more and more educated people to question a literal reading of the Bible. In England, this skepticism contributed to acceptance of Paley's notion of a divine designer, as a means of reconciling nature with the Bible — a concept that was the forerunner of modern-day intelligent design creationism.

---

[1] Darwin, 1859, 450-456.

It was Darwin's theory of evolution that precipitated the resurgence of more widespread creationist beliefs.[1] The concept of biological evolution was readily accepted by many scientists and nonscientists alike once they became aware of Darwin's empirical evidence and his vast collection of biological and fossil specimens, which revealed how much species had changed over the ages. But his proposed evolutionary mechanism of natural selection was more divisive. On both sides of the Atlantic, there were biologists and clergy who could not accept the idea that humans and other creatures evolved from a common ancestor by means of an unguided process with no purpose. This refusal to accept the idea of natural selection led to rejection of the whole theory of evolution by a small band of religious fundamentalists in the early 20th century.

The Protestant fundamentalist movement in America was launched by a series of 12 pamphlets known as *The Fundamentals*, published between 1910 and 1915. Reacting to growing criticism of a literal interpretation of the Bible, especially in Germany, *The Fundamentals* focused on defending often questioned biblical passages. About 20% of the essays touched on evolution but, except for opposition to natural selection, overall treatment of the topic was fairly even-handed. Although fundamentalists did not particularly like evolution, they weren't yet threatened by the concept or by its inclusion in biological textbooks. All this changed drastically after World War I, the death toll and atrocities of which were linked by fundamentalists to German adoption of Darwinism with its misunderstood emphasis on survival of the fittest,[2] although this notion isn't embraced by most historians. At the same time, there was a national conservative backlash in the U.S. that gave rise both to Prohibition and to strident attacks on evolution. Fundamentalists objected especially to the teaching of evolutionary theory in schools.

Leading the fundamentalist movement against evolution was William Jennings Bryan (1860–1925), a Presbyterian and pacifist who had run for President on the Democratic ticket three times without success. Still a powerful figure, Bryan campaigned vigorously in the 1920s for state laws to ban the teaching of evolution. By the end of the decade, five states — Tennessee, Mississippi, Arkansas, Oklahoma and Florida — had outlawed or curtailed teaching the theory of evolution in public schools. Oddly enough, Bryan himself wasn't a biblical literalist but adhered to the so-called day-age theory of creation, in which each "day" of the creation story in Genesis 1

---

[1] Numbers, Ronald L. 1992. *The Creationists*. New York: Alfred A. Knopf, chap. 1.
[2] Gould, Stephen Jay. 2002. *Rocks of Ages: Science and Religion in the Fullness of Life*. New York: Ballantine Books, 155-160.

actually represents a long period or "age" of geological time. Neither did he object to evolution, so long as it excluded humans.

The culmination of creationist efforts to quash evolutionary teaching was the infamous Scopes Monkey Trial of 1925, in which Bryan acted as a prosecuting attorney. Held in Dayton, Tennessee, the trial charged substitute biology teacher John T. Scopes with violating Tennessee's recently passed Butler Act by teaching the theory of evolution in his classes. The Butler Act made it unlawful in state public schools or universities to teach any theory that denied the biblical story of Creation, and to teach in its place "that man has descended from a lower order of animals."[1] Heading the team defending Scopes was legendary Chicago defense attorney Clarence Darrow (1857–1938). Scopes had in fact been put up to incriminating himself by local publicity-seeking town leaders, in response to outreach by the American Civil Liberties Union (ACLU) that was seeking to challenge the Butler Act. But Scopes was happy to oblige and the town soon took on a carnival atmosphere, with thousands of visitors including reporters descending on it. Banners lined the streets, chimpanzees performed in a sideshow and monkey souvenirs were sold.

Once the trial began, Darrow's intended strategy fell apart. The judge refused to allow the defense witnesses, consisting of eight scientists and four religious experts, to testify because their testimony did not address the charge against Scopes of breaking the law, but took issue with the law itself. The judge did, however, permit excerpts from their prepared statements to be read into the record for appeal purposes. But then, in a masterly coup de grace, the defense called the prosecution's Bryan to the witness stand as a Bible expert. Darrow's barrage of cutting questions to Bryan, designed to undermine a literalist interpretation of the Bible, included asking him about a whale swallowing Jonah, Joshua making the sun stand still in the sky, and languages originating from the Tower of Babel. After Bryan stumbled in his answers, Darrow forced him to concede that his acceptance of the Bible as a historical source wasn't as literal as his followers believed. Worst of all, he had to admit in front of a very large crowd gathered at the courthouse that the six days of creation described in Genesis were probably not 24-hour days but much longer periods. Nevertheless, the next day the judge expunged all Bryan's testimony from the record.[2]

The trial itself was inconclusive. Scopes was found guilty as expected and fined $100, but the Tennessee Supreme Court later threw out the case

---

[1] Linder, Douglas. 2018. "Famous Trials: Tennessee Evolution Statutes," http://www. famous-trials.com/scopesmonkey/2128-evolutionstatues. Accessed 11 July 2018.
[2] Ibid. "State v. John Scopes ('The Monkey Trial'): An Account," http://www.famous-trials.com/scopesmonkey/2127-home. Accessed 11 July 2018.

on the technicality that the fine should have been levied by the jury and not the judge. While Darrow had hoped to take the case all the way to the U.S. Supreme Court, Tennessee's anti-evolution Butler law wasn't repealed by the legislature until 1967 — 12 years after a play based on the trial, *Inherit the Wind*, became a Broadway hit.[1]

In the years after the Scopes trial few new anti-evolution laws were enacted, although the existing laws remained on the books. What happened instead is that evolution disappeared from biology textbooks, as publishers succumbed to pressure from fundamentalist teachers and parents who demanded that the topic be excised or at least minimized. This meant that little evolution was taught in U.S. schools for several decades, even though the Scopes trial was considered a victory for evolutionary theory.

But fundamentalists were jolted out of their complacency in the late 1950s following the Soviet Union's successful 1957 launch of Sputnik, the first artificial satellite to orbit Earth. Worried by how far behind the country had fallen in science and technology, the U.S. began to pour billions of dollars into scientific research and science education. Much of the educational effort centered on the updating of science textbooks, which included the resurrection of evolution in biology texts. A new biology text issued in 1963 covered not only Darwin's theory but also Mendelian genetics and molecular biology. Suddenly, hundreds of thousands of U.S. high-school students were made aware for the first time of their apelike ancestors.

Needless to say, the new textbooks expounding Darwinism did not sit well with creationists. In anticipation of a changed public attitude toward both evolution and creationism, Henry Morris (1918–2006) in 1963 founded the Creation Research Society (CRS) and, later in 1972, the Institute for Creation Research as a research offshoot. The CRS succeeded several earlier organizations begun by other fundamentalists in the 1930s and 1940s, which were primarily religious in outlook. The CRS was distinct from its predecessors in that its membership included a substantial number of scientists, drawn from both academic institutions and research laboratories; few scientists had belonged to the earlier societies. Three years after its founding, about 30% of 680 CRS members boasted an advanced degree in science or medicine, with almost 10% being PhDs or MDs — but all were anti-evolutionists, as required by the society's official statement of belief. Most of the PhDs were in biology, though Morris himself was a PhD in hydraulic engineering.[2]

---

[1] Pierce, Kingston J. 2000. "Scopes Trial," *American History Magazine*, August, http://www.historynet.com/scopes-trial.htm. Accessed 11 July 2018.
[2] Numbers, 1992, chap. 11.

The deliberate inclusion of so many trained scientists in the CRS was an attempt to endow creationism with an element of scientific respectability. The society began by commissioning a textbook to supplement and counter the recently published school biology text that had elevated evolution to a level creationists found offensive. But the 15 leading publishers of high-school texts showed no interest at all in the finished CRS product, *Biology: A Search for Order in Complexity*, which was finally produced in 1970 by a Christian publisher. In an effort to present creationism in a favorable scientific light, the book failed to make any case for creationism itself but resorted to what would become a favorite CRS tactic, namely trying to discredit evolution. Few public schools adopted the text and several states banned its use because of its emphasis on religion.

Young-Earth Creationism

One of the topics addressed in *Biology* was so-called flood geology. This is one of the central tenets of the special or young-Earth creationism espoused by the CRS, which holds that humans, animals and plants were created directly by God out of nothing, and that the only biological changes that have taken place since the creation have been within the original created "kinds"; that the earth is between 6,000 and 10,000 years old; and that the planet was reshaped by a massive worldwide flood as described in the biblical story of Noah's ark.

Flood geology stems from the 18th-century theory of catastrophism, which postulated that the earth's geological history had been dominated by a series of catastrophic events such as floods and volcanic eruptions. These catastrophes were invoked by religious geologists of the time to explain the apparent orderly arrangement of animal fossils in rock strata. Fossils that survive the ravages of nature are usually encased in sedimentary rock, which is the most common type of rock on the earth's surface, or as microfossils in deep-sea sediments, and are found in layers, with each layer containing fossils that are distinct from those in higher or lower layers. Sedimentary layers in different geographic locations all show exactly the same sequence of animal or plant fossils, the oldest fossils such as fish and reptiles lying underneath more recent ones such as mammals and birds, with extinct forms preceding living ones. This is known as the principle of faunal succession. Catastrophists identified the sudden breaks in the fossil record with events that completely destroyed all the animals and plants of that epoch, to be replaced by new species in the next epoch. The most recent break they associated with the Genesis flood.

The connection between the fossil record and the great deluge in the Bible story was revived in the early 20th century by Seventh-day Adventist and schoolteacher George McCready Price (1870–1963). Though not a geologist, Price claimed that the biblical flood accounts for the complete fossil record, with most of the animals and plants embedded sequentially in stratified rocks having once existed together before the flood, thus obviating the need for evolution. According to Price, helpless sea creatures and fish rose to the surface and were buried first, while larger land animals and humans fled the floodwaters to higher ground, before being drowned and buried in the topmost strata. Before the sedimentary strata had solidified, the floodwaters carved out natural wonders such as the Grand Canyon and Niagara Gorge. This is how young-Earth creationists compress the history of life into fewer than 10,000 years.[1]

As further ammunition against evolution, Price drew attention to what geologists call overthrusts or thrust faults.[2] Overthrusts result in older rock layers sitting on top of younger ones and, therefore, ancient microfossils appearing above more recent remains such as those of mammals and birds. However, overthrusts are a well-established geological phenomenon worldwide, giving the lie to the claim of Price and other anti-evolutionists that out-of-order fossils prove that the fossil record is meaningless and offers no support to the theory of evolution.

Although Price published two books on flood geology, his ideas weren't taken seriously, even by other creationists outside his Seventh-day Adventist sect. Decades later, Morris and a Princeton-educated coauthor penned *The Genesis Flood*, essentially a repackaged version of Price's earlier theory with additional arguments based on Morris' knowledge of hydraulics. The book, however, made little reference to Price himself, who was regarded by scientists as a crackpot for many of his opinions. Published in 1961 just before the CRS was founded, the tome sold over 200,000 copies in its first 25 years and is still widely read by fundamentalists. Flood geology aims to fit what creationists see as the scientific evidence against evolution, derived from the fossil record, into a literal reading of the Bible, rather than the reverse of interpreting the Bible to conform to the conventional view of evolution.

Nonetheless, *The Genesis Flood* initially drew scathing criticism from day-age creationists, whose old-Earth interpretation of the six days of creation as geological ages, possibly millions of years long, is incompatible with young-Earth beliefs. So too is gap creationism that involves dual creations: an older

---

[1] Numbers, 1992, xi, 72-81.

[2] Overthrusts or thrust faults, which are caused by collisions between tectonic plates, originate when ground on one side of the fault moves upward and over the top of ground on the other side.

creation, followed by a long gap before the original Earth was destroyed and a newer, special creation took place, in which the six creative days were just 24 hours in duration. A literal interpretation of the Bible allows this if the gap occurs between the first two verses of Genesis 1. But despite the fact that radiometric dating techniques for old rocks were well established by the time *The Genesis Flood* appeared, Morris and his fellow CRS members stubbornly ignored the scientific evidence that lent weight to rival forms of creationism — even though the avowed intent of the CRS was to base creationism on science. As a result, many fundamentalists abandoned their beliefs in day-age and gap creationism for young-Earth flood geology.[1]

The widespread adoption of flood geology by fundamentalists in the 1970s marks the beginning of what creationists call "creation science." Relabeling creationism as creation science was largely a public relations move by the CRS, in response to a 1968 ruling by the U.S. Supreme Court that struck down the last of the old state laws banning the teaching of evolution in schools. In the wake of this decision, fundamentalists adopted the new tactic of seeking equal time in science classes for the teaching of both the theory of evolution and creation science, rather than trying to bar evolution from the classroom as they had done earlier in the century. To make creation science more palatable, creationists excluded references to biblical events such as the six days of creation and Noah's ark, while at the same time embracing non-biblical history such as the fossil record. Yet the underlying purpose of creation science was still to promote the special creation described in the Bible.

This creationist venture into the world of science, deceptive though it was as we'll see shortly, alarmed the scientific mainstream. In 1981, the prestigious National Academy of Sciences (NAS) in the U.S. declared that presenting science and religion in the same context results in misunderstanding of both scientific theory and religious belief,[2] a view later echoed and elaborated by American evolutionary biologist Stephen Jay Gould (1941–2002).[3]

A similar stance was taken by the courts when the quest for equal time to teach creation science and evolution in schools was contested. In 1982, a federal judge declared an Arkansas law requiring the "balanced treatment" of creation science and evolution in schools to be a breach of the U.S. constitutional barrier between church and state. In 1987, the U.S. Supreme Court upheld a ruling by a Louisiana judge that creation science served a religious, not scientific, purpose and that the state's equal-time law was also

---

[1] Numbers, 1992, chap. 10.

[2] Ibid, 241-251.

[3] Gould, 1999, chap. 2.

unconstitutional.[1] Ironically, at the Arkansas trial, all the religious witnesses who testified opposed the teaching of creation science, and no religious groups supported it; most of the defense witnesses supporting creation science were scientists and not religious figures.

Intelligent Design

Faced with these legal defeats for creation science, creationists in the 1990s came up with yet another strategy: the theory of intelligent design (ID), which harks back to Paley's watchmaker argument in the 18th century. ID creationism is distinctly different from the creation science promoted by the CRS. Its central thesis is that complexity in nature implies that the natural world was designed by an intelligent designer, that it could not have originated in any other way. But unlike Paley's watchmaker analogy or creation science, ID does not identify God as the designer or creator — a clever ploy on the part of creationists, who still want to overthrow evolution but without any direct connection to the Bible. The designer, although implicitly assumed to be God, could also be an alien or other supernatural being. Also at odds with creation science is acceptance by most ID proponents of the antiquity of life on earth, in contrast to the 10,000-year maximum allowed by young-Earth creationists.

The ID movement began with the publication of several books in the 1980s, including a supplement to U.S. high-school biology texts that by then had included evolution and genetics for more than 25 years. But it wasn't until 1991 that ID made a splash, when University of California, Berkeley law professor Phillip Johnson published *Darwin on Trial*. His critical examination of the evidence for evolution and natural selection was followed in 1995 by *Darwin's Black Box: The Biochemical Challenge to Evolution*, a book by Lehigh University biochemist Michael Behe. Behe argued that the "astonishing complexity" of the living cell was unambiguous evidence of an intelligent designer at work.[2]

Behe introduced the concept of irreducible complexity, illustrating his idea with the example of the household mousetrap. The mousetrap, he claims, ceases to function and catch mice if any of its five basic components — platform, metal hammer, wire spring, catch and holding bar — is removed. He cites this as an irreducibly complex system because the mousetrap doesn't

---

[1] The justices dissenting from the Supreme Court's seven to two majority opinion were Chief Justice William Rehnquist and Justice Antonin Scalia.

[2] Numbers, Ronald L. 2007. "Darwinism, Creationism, and 'Intelligent Design'." In Andrew J. Petto and Laurie R. Godfrey, eds., *Scientists Confront Intelligent Design and Creationism*, New York: W. W. Norton & Company, 48-54.

work until all five components are assembled as a unit. In Behe's view, an irreducibly complex biological system which involves multiple biochemical pathways can't possibly evolve via the step-by-step mechanism of natural selection or the gradual modification of a precursor system; such a complex system can only be formed as an integrated whole and must therefore have been designed by an intelligent entity.[1] Scientists quickly demolished Behe's original claim, demonstrating that a functional mousetrap can indeed be constructed from as few as one of its five components.[2] Others pointed out that natural selection can work on modular units already assembled for another purpose, so the components for a complete mousetrap need not be created from scratch, reducing the total number of steps.

Another intelligent design concept is that of complex specified information (CSI) or specified complexity. According to U.S. mathematician and philosopher William Dembski, the detection of CSI in a biological system is irrefutable proof of ID. While Dembski has given elaborate mathematical explanations of "complex" and "specified," he has also stated more simply that:

> A single letter of the alphabet is specified without being complex. A long sentence of random letters is complex without being specified. A Shakespearean sonnet is both complex and specified.[3]

Specified denotes a match to a pattern, independent of the information itself. To ascertain whether a particular event or particular biological system exhibits specified complexity, Dembski uses what he calls an explanatory filter that separates events or systems into three categories: high, intermediate and low probability. High-probability events, such as the rising of the sun, are explained by natural laws; intermediate- and unspecified low-probability events are attributed to chance; but specified low-probability occurrences supposedly can only result from design. Dembski claims to find CSI in DNA and in biological complexes such as the bacterial flagellum,[4] implying that these systems have an intelligent designer.[5] But the notions

---

[1] Behe, Michael J. 2001. "Molecular Machines: Experimental Support for the Design Inference." In Robert T. Pennock, ed., *Intelligent Design Creationism and its Critics: Philosophical, Theological, and Scientific Perspectives*, Cambridge, MA: The MIT Press, chap. 10.

[2] McDonald, John H. 2011. "A Reducibly Complex Mousetrap," 14 March, http://udel.edu/~mcdonald/mousetrap.html. Accessed 11 July 2018.

[3] Faith from Evidence. "What is Intelligent Design?," http://www.faithfromevidence.org/what-is-intelligent-design.html. Accessed 11 July 2018.

[4] The bacterial flagellum is a whip-like appendage on a bacterial cell that rotates like a propeller, enabling bacteria to swim rapidly through host fluids or across surfaces. Some bacteria have multiple flagella.

[5] Dembski, William A. 2001. "Intelligent Design as a Theory of Information." In Robert T. Pennock, ed., *Intelligent Design Creationism and its Critics: Philosophical, Theological, and Scientific*

of CSI and irreducible complexity have both been heavily criticized, a topic we'll come back to in the next section.

Not surprisingly, like young-Earth creationism before it, ID found its way into U.S. courts. In 2004, the school board in Dover, Pennsylvania angered teachers and some parents by opening the door to teaching ID in high-school biology classes, with the resolution: "Students will be made aware of gaps/problems in Darwin's theory and of other theories of evolution including, but not limited to, intelligent design." The creationist-leaning board also issued a press release requiring biology teachers to draw attention to an ID reference book, *Of Pandas and People*, 60 copies of which had been donated to the school and placed in the library.[1]

When the school's teachers refused to read the requisite statement about the donated book to their classes, administrators read it instead, prompting a lawsuit against the school district by parents opposed to the new policy. The plaintiffs were represented in court by the ACLU, reprising its role in the Scopes trial 80 years before, and by another civil liberties organization. The case went to trial in 2005, charging that the board's resolution and press release sanctioned the teaching of religion in a public school and were therefore unconstitutional. For ID, this was the same accusation as the charge made against creation science in Louisiana that the U.S. Supreme Court had upheld in 1987. The defense in the Dover trial therefore hinged on convincing the presiding judge that ID is a science and not a form of creationism, a task in which it failed miserably.

After several weeks of testimony, the judge ruled that the school board's policy violated the constitution because ID was no more scientific than its predecessor, creation science. In fact, the judge was "astonished" when testimony revealed that in early drafts of *Of Pandas and People*, the words "creationism" and "creationist" had been replaced approximately 150 times with the phrase "intelligent design," the switch tellingly coinciding with the 1987 Supreme Court decision.[2] The judge also castigated the board members for not knowing "precisely what ID is," and concluded that some of them had lied under oath, both in depositions and on the witness stand.

After ID suffered such a major defeat in the Dover trial, its proponents turned away from trying to introduce ID into the classroom. Instead, the

Perspectives, Cambridge, MA: The MIT Press, chap. 25.

[1] The TalkOrigins Archive. 2005. "Kitzmiller v. Dover Area School District: Decision of the Court, Introduction," 31 December, http://www.talkorigins.org/faqs/dover/kitzmiller_v_dover_decision.html. Accessed 11 July 2018.

[2] The TalkOrigins Archive. 2005. "Kitzmiller v. Dover Area School District: Decision of the Court, E. Application of the Endorsement Test to the ID Policy, 1.," 31 December, http://www.talkorigins.org/faqs/dover/kitzmiller_v_dover_decision.html. Accessed 11 July 2018.

well-funded Discovery Institute, which plays much the same role for ID as the CRS does for young-Earth creationism, has emphasized teaching what it sees as the deficiencies of the theory of evolution, along with its merits — a similar tactic to that used by the CRS for many years.

In spite of the legal setbacks to teaching creationism in schools, anti-evolutionism still has a large following in the U.S. According to a survey conducted by the Pew Research Center in 2014, the percentage of the U.S. population overall that believes humans have always existed in their present form, without evolution, is 34%, and the percentage is even higher for those who identify with certain Christian faiths and for Muslims. The 34% is a measure of young-Earth creationists; ID proponents are included in the 25% who believe in divinely guided evolution.[1] A 2017 Gallup poll noted a doubling in America since 1999 of the percentage of strict evolutionists, who believe in evolution free of God's hand, with creationists representing 38% of the population.[2]

Globally, the percentage of people who regard themselves as creationists was 37% in 2011, about the same as in the U.S.[3] This shows the breadth of disbelief in Darwin's celebrated theory, even though it is almost universally accepted by the world's scientists and mainstream churches. Significant creationist movements now exist in many countries including Canada, Australia, the UK, South Korea and even the Islamic nation of Turkey.[4]

Why Creation Science And Intelligent Design Aren't Science

As we've seen in earlier chapters, the distinguishing characteristics of the scientific method are empirical evidence, hypothesis and reason. In this chapter, we've touched on the overwhelming evidence for the theory of evolution: the fossils and living species that Darwin collected on his *Beagle* voyage, together with the subsequent abundance of modern evidence from fields such as molecular biology and embryology. The logic used to make sense of all these observations, along with Darwin's hypothesis of natural

---

[1] Pew Research Center. 2015. "U.S. Public Becoming Less Religious. Chapter 4: Social and Political Attitudes," 3 November, http://www.pewforum.org/2015/11/03/chapter-4-social-and-political-attitudes/. Accessed 11 July 2018.

[2] Swift, Art. 2017. "In U.S., Belief in Creationist View of Humans at New Low." *Gallup*, 22 May, https://news.gallup.com/poll/210956/belief-creationist-view-humans-new-low.aspx. Accessed 11 July 2018.

[3] Ipsos Polls. 2011. "Ipsos Global @dvisory: Supreme Being(s), the Afterlife and Evolution," 3 May, http://marketresearchworld.net/content/view/4022/77/. Accessed 11 July 2018.

[4] Numbers, 2007, 42-43.

selection, elevate the evidence to a full-fledged theory. There's no question that the theory is science, both in terms of the literal meaning of science as knowledge of nature, and in terms of adhering to the scientific method. Even though the fossil record is incomplete and we have yet to find certain transitional forms, we have already found more than enough to demonstrate the soundness of evolutionary theory.[1]

On the other hand, creationism can't be called science by any stretch of the imagination — a conclusion that the judge reached in the Dover trial on teaching ID in biology classes, and discussed at length in his 139-page opinion.[2] The opinion, while a legal document, is heavily based on scientific evidence presented at the trial, and cites the primary reason that ID is not science as its reliance on a supernatural designer. Science does not embrace supernatural causes of phenomena in the natural world, which the U.S. Supreme Court in 1987 recognized as an inherently religious concept in its ruling against so-called creation science.

The law and religion aside, however, there are many good scientific reasons why creationism doesn't qualify as science. For a start, creation science and ID either distort or ignore the vast body of evidence that constitutes the fossil record, simply to reject the theory of evolution. And creationists are hard-pressed to provide an alternative explanation of the evidence for evolution supplied by biogeography and DNA sequencing. For example, the presence of marsupial mammals in Australia, independent of placental mammals elsewhere on the globe, is inexplicable by creationism without making all sorts of unreasonable assumptions — such as created marsupials getting to Australia ahead of created placentals, but not to other continents, before Australia drifted apart and isolated the marsupials. Likewise, creationism can't explain the existence of many of the same dead genes in all members of the same species, such as humans and chimpanzees among the primates.

The two areas in which creationists claim that their religious beliefs are underpinned by science are the flood geology of young-Earth creationists, and the concepts of irreducible complexity and specified complexity in ID. Apart from the fact that no scientific evidence exists for the global flood described in Genesis 6–8, the authors of the creationist book *The Genesis Flood* have deluded themselves that the fossil record is always as neatly ordered as

[1] Prothero, Donald R. 2007. *Evolution: What the Fossils Say and Why It Matters.* New York: Columbia University Press, 125-126, 188-195.
[2] The TalkOrigins Archive. 2005. "Kitzmiller v. Dover Area School District: Decision of the Court, E. Application of the Endorsement Test to the ID Policy, 4. Whether ID is Science," 31 December, http://www.talkorigins.org/faqs/dover/kitzmiller_v_dover_decision2.html. Accessed 11 July 2018.

shown in basic biology or geology texts, and that such a flood could account for the geological features of the Grand Canyon and other works of nature.

In *The Genesis Flood*, Morris and his coauthor augmented the earlier flood geology model of Price with the notion of hydraulic sorting. Their explanation of the fossil record is that marine creatures and fish were buried first by the rising waters as in Price's model; more sophisticated animals such as amphibians and reptiles (including dinosaurs) were able to flee to intermediate levels; and the smarter mammals climbed to the highest levels before being drowned and buried. This scenario would produce a fossil sequence just like the so-called geologic column presented in textbooks. The geologic column is an idealized, conceptual representation of the sedimentary rock layers laid down during different periods of the earth's history, with accompanying fossils from the same period, conforming to the principle of faunal succession whereby fossils accumulate in a specific order.[1] The flood geology picture, however, is far from reality and has little in common with most known fossil sequences on the planet.

One of several reasons that the simple-minded flood geology picture is a fantasy is that the same rock layers can exist at completely different altitudes in different geographic locations. As an example, well-preserved fossils of fish and other small animals are found in the Green River Formation of lake sediments in Utah in the U.S., where both the fossils and sediments are typical of deposits in lakes that dried up periodically, but not of a huge flood. And the fish fossils lie in height *above* the dinosaur fossil beds at the Dinosaur National Monument on the Colorado–Utah border. The burial of fish at a higher altitude than the supposedly faster dinosaurs can in no way be reconciled with flood geology, any more than the tortoise fossils overlying the fossilized mammals in South Dakota's Brule Formation can, although both these fossil discoveries are consistent with the geologic column. It's ludicrous that laggardly tortoises somehow outran horses and rhinos in fleeing the floodwaters. Numerous other fossil examples all over the globe also contradict the flood model.

Another reason that flood geology is preposterous is that certain organisms such as hard-shelled mollusks — living species include snails, squid, octopi,

---

[1] Creationists often claim that the geologic column involves circular reasoning, with rocks being dated by their fossil content and fossils being dated simultaneously by their position in the geologic column. This is a fallacy, since unknown fossils can be dated approximately through comparison with other fossils found in the same rock layer, using the principle of faunal succession. More precise dating of either fossils or rocks usually involves radiometric methods.

The distribution of fossils in different strata, known as biostratigraphy, is utilized extensively in the oil and gas industries to date and correlate rock layers encountered during drilling.

clams and oysters — appear throughout the fossil record, so are found in multiple strata, the specific strata depending on the local environment. This again is incompatible with a massive flood that would have drowned mollusks at the bottom level. Likewise, no one has ever discovered human and dinosaur fossils entombed together in the same upper strata, though this is a necessary consequence of flood geology. Because dinosaurs disappeared from the earth 60 million years before the first humans were here, their fossils only turn up in much lower sedimentary layers than human remnants.

As for the Grand Canyon, the creationist explanation of its features is nullified by the same total lack of evidence as that for the flood geology explanation of the fossil record. The spectacular canyon is one of the world's most complete geologic columns, with nearly 40 sedimentary rock layers up to 1.8 billion years old. If the Grand Canyon had truly been sculpted by the great biblical flood, it would look utterly different. For example, the canyon walls, not to mention other rocky areas of the world, would be covered with a thin, single layer of mudstone, resulting from the mud that over time settles out of suspension from standing floodwaters. The closest that creationists can come to explaining the absence of this enveloping mud layer is by pointing to the various mudstone and shale layers in the Grand Canyon's geologic sequence. Many of these layers contain hundreds of mud-cracked sublayers, some full of tracks and burrows made by tiny animals, that signify multiple episodes of mud deposition and subsequent drying out — but not a single catastrophic flood that would have quickly drowned all the creatures before they had a chance to burrow. Creationists ignore evidence and throw reason out the window.

Among other Grand Canyon quirks inexplicable by flood geology is a series of ancient rocks at the base that are tilted on their sides from some long-ago tectonic event. The rest of the canyon strata lie horizontally on top of them. Had the tilted rock layers been deposited as soil layers from a major flood, they would have all folded over under the influence of gravity once the tilting occurred. But what geologists actually find is an undisturbed sequence of tilted layers, replete with ancient microfossils, mud cracks and volcanic lava flows. Needless to say, neither the authors of *The Genesis Flood* nor Price had any background in geology.

Although unscientific flood geology is the only basis of creation science, young-Earth creationists are just as unscientific in attempting to justify their belief in an earth 6,000 to 10,000 years old as they are in invoking the great biblical flood to explain the fossil record. The actual age of the earth is 4.5 to 4.6 billion years, ascertained by well-developed radiometric dating methods. Although it wasn't established until about 90 years ago that the planet was at least a billion years old, even that lower estimate was a far cry from the several thousand years that young-Earth creationists allow, derived from a literal reading of the Bible.

To prop up their young-Earth belief, creationists have come up with a slew of fallacious claims about radiometric dating, not unlike their efforts to cast doubt on evolution. One claim is that the method is unreliable because of a tiny number of instances when radiometric dating has been incorrect, an argument that doesn't prove a young earth, and in any case doesn't invalidate a technique that has yielded correct results tens of thousands of times. Another, equally ridiculous claim is that somehow the *rate* of radioactive decay, on which the dating technique depends, was billions of times greater in the past, which would considerably shorten radiometrically measured ages. Some creationists even claim that radioactive decay sped up more than once. However, any substantial change in decay rates would require changes in fundamental physical constants (such as the speed of light), the alteration of which would mean we'd be living in a completely different type of universe.[1] Among other wild assertions that creationists use as evidence that the planet is no more than 10,000 years old are rapid exponential decay of the earth's magnetic field, which is a fictitious claim,[2] and the low level of helium in the atmosphere, which merely reflects how easily the gas escapes from the earth and has nothing to do with its age.

So despite all the deceitful efforts of the CRS to present creationism as science, it most definitely isn't. As U.S. courts have ruled, creationism is a religious belief, based on faith in a divine creator and a literal interpretation of the Bible. There's nothing wrong with faith, but it's the very antithesis of science. Science requires evidence and a skeptical evaluation of claims, while faith demands unquestioning belief, without evidence. It takes faith on the part of creationists to believe in sudden creation of the present world, and in a young earth and flood geology, especially when solid scientific evidence exists refuting all of these — just as it takes faith for the religious to believe in a God, although the existence of God can't even be tested scientifically. Creationists go to absurd lengths attempting to contort the facts to fit their scientific preconceptions, which are based on nothing more than their reading of Scripture.

Turning to ID, we've already seen how Behe's mousetrap example of irreducible complexity has been invalidated. A household mousetrap can

---

[1] Dalrymple, G. Brent, 2007. "The Ages of the Earth, Solar System, Galaxy, and Universe." In Andrew J. Petto and Laurie R. Godfrey, eds., *Scientists Confront Intelligent Design and Creationism*, New York: W. W. Norton & Company, 150-173.

[2] Although the earth's magnetic field has been decreasing slowly for at least 200 years, the drop so far is within the field's historical range of variation.

still perform its function with only a single component, albeit less efficiently than one with all five components which, therefore, isn't irreducibly complex. In nature too, it's a fallacy that a complex biochemical system can function only if all its components have been assembled as a module, and not step by step through the natural selection process. Contrary to what Behe claims, complex biochemical structures or processes such as the wondrously intricate human blood clotting cascade can indeed be built up from simpler systems by natural selection. The blood clotting mechanism, like all biochemical systems, evolved from proteins and genes that originally served entirely different purposes, having nothing to do with clotting. For example, an enzyme that now plays a central role in blood clotting, by clipping off a protective barrier from clotting proteins to expose the sticky part, was once employed for signaling between tissue cells. During hundreds of millions of years of evolution, natural selection favored the genetic mutations that resulted in this enzyme change beneficial to blood clotting.[1]

Even biological systems that truly are irreducibly complex, where removal of just one component would cause the system to cease operating, can be formed through step-by-step evolution, by a process known as scaffolding. This refers to the building of a structure such as a stone arch that is built stone by stone, with supporting scaffolding to hold the sides of the arch in place during construction. After the crucial keystone is laid at the top, the scaffolding is removed since the irreducibly complex arch can now stand on its own. Similarly, evolution can produce an irreducibly complex biochemical structure, when natural selection works to remove unnecessary components.

Both irreducible complexity and CSI or specified complexity are examples of what philosophers call the argument from ignorance. In the case of ID, the argument is that certain biological systems such as the bacterial flagellum are so highly complex as to defy scientific explanation, and therefore must have been designed. In other words, if we can't think of any way that nature or chance could have produced the flagellum, an intelligent designer must have created it. As discussed before, ID is careful not to name the designer, although ID creationists will quickly admit they see the designer as God. Nevertheless, a perfectly adequate scientific explanation exists for the origin of the bacterial flagellum, in the form of the theory of evolution, just as it does for the blood clotting cascade. And if there weren't an explanation today, that doesn't mean science wouldn't find one tomorrow. There's no scientific evidence at all for design, nor any logic that supports the faith-

---

[1] Miller, Kenneth R. 2005. "The Evolution of Vertebrate Blood Clotting," http://www.millerandlevine.com/km/evol/DI/clot/Clotting.html. Accessed 11 July 2018.

based concept. Dembski's explanatory filter for supposedly identifying design is no more than a formalization of the classic God-of-the-gaps fallacy — a theological variation of the argument from ignorance, in which gaps in scientific knowledge are attributed to acts of God.

CSI and irreducible complexity are both quasi-scientific attempts to legitimize ID, akin to the use of flood geology to validate young-Earth creationism. But neither constitutes a valid scientific alternative to evolution, in spite of Dembski and Behe claiming just the opposite. An example of complexity frequently cited by ID proponents is the human eye, an incredibly complex organ that Darwin himself recognized as a serious challenge to his theory. But what ID believers miss in seizing on Darwin's rhetorical statement about the "extreme perfection" of the eye, and the difficulty in attributing it to natural selection, is that he went on to say that reason told him the difficulty wasn't insurmountable, well before it was even understood how natural selection actually works.[1] Both ID proponents and young-Earth creationists fail to appreciate the enormous lengths of time required for natural selection to evolve biological structures as sophisticated as the eye. Because eyes don't fossilize, we don't know how long it took the human eye to evolve to its present complexity. Darwin guessed that it would be at least the age of the oldest fossils, or several hundred million years, which is more than several million lifetimes and intuitively meaningless to us. But that's a lot of generations for natural selection to accomplish its incremental improvements to the primitive eye.

Apart from not offering any scientific evidence against evolution, neither creation science nor ID reveals the foresight and sense of purpose that we would expect from the type of designer envisaged by Paley. In nature, there are numerous instances of poor or inefficient design, where some body part in an animal has been jury-rigged from another, totally unrelated part. An often cited case is the female deep-sea anglerfish that has a thin "fishing pole" with a luminous tip protruding from its forehead, used to lure prey close enough to suck into its mouth; the fishing pole evolved from a dorsal fin spine. Another example is the panda's "thumb," an addition to the five fingers of its paw that is just strong enough to allow a panda to strip the leaves off bamboo shoots, on which the species subsists; the thumb evolved from a wrist bone. Both cases show clumsy, jury-rigged design, certainly not intelligent design, but at least design that works for the task at hand. Good enough to ensure survival is all that natural selection needs to provide.

That creation science and ID aren't science at all is reflected in the almost total lack of peer-reviewed papers published in the scientific literature. Aside

---

[1] Darwin, 1859, 186-189.

from the occasional article in educational journals on the different forms of creationism,[1] the only known paper on creationism itself — an ID article on the so-called Cambrian explosion — appeared in an obscure biological journal in 2004. But a month later, the journal's publishing society chastised the editor for improperly handling the peer review process and repudiated the article, adding that there was no scientific evidence supporting ID,[2] as I've just discussed. Scientific peer review was abused in this case for the sake of religious beliefs; we'll see in later chapters how it gets abused for political and other reasons. Because there's no science in creationism, its advocates rarely attend scientific meetings and resort to publishing their own books and journals to promulgate their views.

In the face of an onslaught of scientific criticism, both young-Earth creationists and ID proponents respond the same way, by rejecting the theory of evolution just as early American fundamentalists did. But while refusing to accept that evolution has a perfectly sound explanation of the fossil record, diversity and vestigial structures, creationism offers no scientific alternative to Darwin's hypothesis of natural selection. Flood geology, irreducible complexity and CSI are all pseudoscience because they depend not on scientific evidence, but on religious beliefs and therefore can't be falsified, as required of a valid scientific theory. Instead of proposing their own scientific theory to supplant evolution, creationists rely on trying to poke holes in the less than 1% of evidence from the fossil record that can't currently be explained, or by finding fault in the molecular models put forward to explain unknown details of complex biochemical processes — and then falling back on the argument from ignorance to invoke design as the answer. But these efforts fall far short of overturning evolutionary theory.

At root, creationism is, and always has been, about religion and a belief that God created the natural world in essentially its present form. This faith-based belief is incompatible with the scientific theory of evolution, a conflict that Darwin himself wrestled with in an era when creationist views were common. Today, however, many scientists accept evolution yet have a faith at the same time. Some choose to believe that the evolutionary process is guided by God, so-called theistic evolution that became popular in the late 19th century; others that include me believe that God simply created

---

[1] See, for example, Ross, Marcus R. 2005. "Who Believes What? Clearing up Confusion over Intelligent Design and Young-Earth Creationism." *Journal of Geoscience Education*, 53, 319-323.

[2] Statement from the Council of the Biological Society of Washington. 2007. https://web.archive.org/web/20070926214521/http://www.biolsocwash.org/id_statement.html. Accessed 11 July 2018.

the laws of nature, including the principle of natural selection, but plays no other role in evolution; and some scientists who accept evolution are agnostics or atheists. As Gould discussed at length, science and religion are completely separate domains. Reason alone demands that faith in the form of creationism should not be allowed to masquerade as science.

By pretending to be science, creationism is muddling the public perception of science, fostering pseudoscience and undermining the scientific method, on which so much of our modern world has been built.

Chapter 4. Dietary Fat: Nutritional Politics

Two of the most visible areas of science in the public sphere are medicine and nutrition science. Hardly a month goes by without the announcement of some astonishing breakthrough in medical science or the rollout of some newfangled diet. In this chapter we'll examine the once much vaunted link between dietary fat and heart disease, a connection now in considerable doubt as the result of recent analyses and reviews of data gathered over the last 60 years. Yet the association of coronary heart disease (CHD) and deaths with what we eat is still deeply entrenched in the public consciousness, at least in the U.S. and UK.

This is the first of several examples in the book of how science today is being subverted by political forces that overlook evidence and sideline reason, here in the worlds of nutrition and public health. The intrusion of politics into nutritional science dates back to the introduction of official government dietary guidelines, in 1980 in the U.S. and 1983 in the UK. However, while dietary fat dogma has been perpetuated by the power of a new nutritional establishment that controls research agendas and government funding, the origin of the nutritional dogma itself wasn't political but rather the sheer force of personality of one man, the American physiologist Ancel Keys (1904-2004).

The Dietary Fat Story

The so-called diet–heart paradigm — the hypothesis that saturated fat is the demon behind CHD — was conceived by Keys in the 1950s, although a linkage between dietary cholesterol and CHD had been suggested much earlier. At the time, there was widespread alarm at the surge in CHD, which is caused by narrowing of the coronary arteries, since the beginning of the 20th century. Although CHD is the predominant type of heart disease and can lead to a heart attack,[1] the medical community in the 1950s largely believed that narrowing of the arteries was a result of aging and couldn't be avoided, any more than gray hair or wrinkles. However, Keys thought that heart attacks could be prevented and that dietary fat was implicated. He was already famous for developing the K-ration, an emergency meal ration distributed to combat soldiers in World War II.

The prevailing wisdom when Keys first took an interest in CHD was that the condition resulted from excessive ingestion of cholesterol, the waxy, fat-like substance that is also manufactured by the liver and is an essential part of our cell membranes. The association of cholesterol with CHD came from misinterpretation of the symptoms of a rare genetic disease in children, as well as pioneering laboratory experiments reported in 1913 by Russian research pathologist Nikolai Anichkov (1885–1964), in which rabbits were fed enormous quantities of cholesterol.[2] The rabbits subsequently developed atherosclerosis, the term used to describe the thickening and constriction of arteries due to the accretion of plaque in the artery walls, and Anichkov connected the plaque with the excess cholesterol in the animals' diet.

Although this research was later criticized because herbivorous rabbits don't readily metabolize cholesterol as humans do, other researchers found that cholesterol was indeed a major component of human atherosclerotic plaque and that cholesterol in plaque had the same composition as that in the bloodstream, known as serum cholesterol.[3, 4] While plaque accumulation itself does not clog the arteries as once thought, it does cause the arteries to narrow, which can lead to both CHD and several forms of cerebrovascular disease in the brain.[5]

---

[1] Apart from coronary heart disease (CHD), sometimes called coronary artery disease, other types of heart ailment include abnormal heart rhythms (arrhythmias), congenital heart defects, enlargement of the heart muscle (cardiomyopathy), and heart infection. A heart attack is often referred to as coronary thrombosis or myocardial infarction.

[2] Anitschkow, N. and S. Chalatow (Anitschkow is the German spelling of the Russian Anichkov). 1913. "Classics in Arteriosclerosis Research: On Experimental Cholesterin Steatosis and its Significance in the Origin of Some Pathological Processes," translated by Mary Z. Pelias. *Arteriosclerosis, Thrombosis, and Vascular Biology*, 3, 178-182.

[3] Hirsch, Edwin F. and Sidney Weinhouse. 1943. "The Rôle of the Lipids in Atherosclerosis." *Physiological Reviews*, 23, 185-202.

[4] Gofman, John W., Frank Lindgren, Harold Elliott et al. 1950. "The Role of Lipids and Lipoproteins in Atherosclerosis." *Science*, 111, 166-186.

[5] Cholesterol does not dissolve in the blood, but is encased in lipoproteins that carry triglyceride fats and cholesterol through the body's arteries and veins. The principal lipoproteins are high-density lipoprotein (HDL) and low-density lipoprotein (LDL). HDL-cholesterol is known as "good" cholesterol, LDL-cholesterol as "bad" cholesterol;

In a diet study reported by Keys in 1953, he discovered that serum cholesterol appeared to be unrelated to how much cholesterol the study volunteers were fed, a finding that was later questioned. But in the same study, Keys first linked CHD to fat in the diet, rather than cholesterol.[1] This formal introduction of his diet–heart hypothesis arose from statistical data that he had collected from six countries, for men in two age groups (45-49 and 55-59) who had died from CHD. His graph of death rate against fat intake revealed a clear correlation between the two, deaths rising smoothly with the percentage of total calories from fat, in both age groups. The graph was the genesis of the idea that dietary fat causes heart disease — an idea that mushroomed and led to government intervention in the dietary arena, as well as spawning the huge low-fat industry.

Keys, convinced he was right, was a tireless promoter in the nutrition community of his belief that CHD resulted primarily from "the long-time effects of a rich fatty diet and innumerable fat-loading meals."[2] However, he had the wind at least temporarily taken out of his sails after a presentation at a 1955 conference in Geneva. University of California, Berkeley biostatistician Jacob Yerushalmy (1904–73) and New York State Commissioner of Health Herman Hilleboe (1906–74), who had both been irritated by Keys' Geneva talk, complained in a 1957 paper that Keys had presented as "a proved fact" his assertion that there was "a strong association between the mortality from heart disease and proportion of fat available in the diet in different countries," but the assertion was based on only a handful of countries.[3]

Yerushalmy and Hilleboe put together their own statistical data from not six but 22 countries, for men aged 55 to 59, that still showed a correlation between CHD mortality and fat consumption, but the correlation was much weaker than Keys had found. Their 22 countries included several in Europe such as France, West Germany and Sweden, whose populations indulged in relatively large amounts of fat but died less frequently from CHD on average. They also found that the association of dietary fat with CHD was no stronger

the two terms refer to the effect of that type of cholesterol on atherosclerosis. Atherosclerosis involves the buildup of plaque consisting of calcium and oxidized LDL cholesterol inside an artery wall, causing the artery to narrow over time. If the plaque ruptures into the bloodstream, blockages can occur, triggering a heart attack or ischemic stroke.

[1] Keys, Ancel. 1953. "Atherosclerosis: A Problem in Newer Public Health." *Journal of the Mount Sinai Hospital, New York*, 20, 118-39, http://www.wisenutritioncoaching.com.au/wp-content/uploads/2013/07/Keys-Atherosclerosis-A-Problem-in-Newer-Public-Health.pdf. Accessed 11 July 2018.

[2] Keys, Ancel. 1957. "Diet and the Epidemiology of Coronary Heart Disease." *Journal of the American Medical Association*, 164, 1912-1919.

[3] Yerushalmy, J. and H. E. Hilleboe. 1957. "Fat in the Diet and Mortality from Heart Disease; a Methodologic Note." *New York State Journal of Medicine*, 57, 2343-2354.

than for dietary protein with CHD. However, what correlated with CHD was *animal* fat and protein; the correlations for plant fat and protein were negative — a prelude to Keys' later focus on saturated fats, the majority of which come from animal products. Nonetheless, Yerushalmy and Hilleboe ignored much of Keys' statistical evidence and patronizingly implied that his selection of countries was biased, rubbing in the fact that he had only shown an *association* between fat and heart disease, not *causation*. Of course their own statistical study was subject to the same limitation.

Chastened by this experience, Keys decided to expand trial surveys that his small team had embarked on, in several countries, into a major epidemiological study that became known as the Seven Countries Study. International statistical studies of the type Keys, and Yerushalmy and Hilleboe, had already conducted are notoriously unreliable, because of varying record-keeping procedures and vastly different methods of diagnosing the cause of death from country to country. An epidemiological or so-called observational study is intermediate between a purely statistical survey and a clinical or randomized controlled trial. Epidemiology seeks connections between death from a disease and various measurable factors identified as possible causes of the disease; the actual cause can only be inferred, nonetheless, just as in a statistical investigation. In Keys' study, the measurements included medical data such as weight, blood pressure, serum cholesterol and smoking history, together with dietary factors such as saturated fat consumption.

The Seven Countries Study compared the dependence of CHD on diet and other lifestyle factors in 12,770 middle-aged men between 40 and 59 in Finland, Greece, Italy, Japan, the Netherlands, the U.S. and what was then Yugoslavia. The first multicountry epidemiological study and one of the longest such studies ever, it followed its subjects for up to 40 years, with reexamination of survivors every five or ten years.

Started over the period from 1958 to 1964, the study reported its initial five-year follow-up results in 1970. These results showed that across the seven countries, CHD incidence and mortality were both strongly correlated with serum cholesterol as well as the percentage of calories from saturated fat in the diet, but not with the percentage of calories from all fats[1] — in contrast to Key's earlier six-country statistical study. The later 15-year follow-up discovered surprisingly that CHD accounted for less than a third

---

[1] Keys, Ancel (ed.). 1970. "Coronary Heart Disease in Seven Countries." *Circulation*, 41, Supplement 4S1, published as American Heart Association Monograph No. 29, April 1970, I-162 - I-183.

of deaths from all causes,[1] an issue we'll return to later. As in the first five years, CHD deaths after 15 years correlated with serum cholesterol, though the overall death rate did not correlate. The results also showed a large spread from country to country, with CHD causing 46% of all study deaths in the U.S., but only 17% in southern Europe and fewer than 10% in Japan.

Although Keys, nicknamed "Mr. Cholesterol" by some of his colleagues, went to great lengths to choose stable study populations and to standardize measurement procedures suitable for a wide variety of cultures and for groups in remote areas, the Seven Countries Study can be faulted on several grounds. Probably its biggest limitation was the somewhat arbitrary selection of countries and of population groups within countries, both choices being made at least partly for convenience, though funding shortages played a role too.

Even though trial surveys had already been conducted in Spain and South Africa, Spain was excluded from the main study supposedly due to lack of local interest. South Africa was omitted not because coronary disease was rare as in Japan, which was included, but because of the "unfavorable status" and lack of enthusiasm for "imitating the ways" of the Bantu people.[2] As in Keys' six-country statistical survey, France and West Germany, both of which had highly fatty diets but few deaths from CHD, weren't included. Nor was Chile, where fat consumption was low but CHD mortality high; Chile, along with France and West Germany, were part of Yerushalmy and Hilleboe's 22-country statistical survey, however. Six of Keys' seven countries exemplified high fat consumption combined with a high CHD death rate, or low fat consumption combined with a low CHD death rate. Nevertheless, he did include the two statistical extremes of Japan, that has one of the lowest CHD death rates, and Finland, that has one of the highest — which Keys suspected was related to diet, having previously observed that the muscular farmers and loggers in Finland like to butter their cheese.

Another shortcoming of the Seven Countries Study was in its collection of dietary data. With the exception of the U.S., diets in each of the 16 different study cohorts were sampled in seven-day surveys, during which all food eaten by the men in that sample was weighed and analyzed, this being repeated in say summer and winter to account for seasonal variations. The U.S. cohort was instead required to complete a two-page dietary questionnaire detailing what the men had eaten the previous day, with a tiny

---

[1] Keys, Ancel, Alessandro Menotti, Christ Aravanis et al. 1984. "The Seven Countries Study: 2,289 Deaths in 15 Years." *Preventive Medicine*, 13, 141-154.

[2] Keys, 1970, I-2 - I-3.

sample being interviewed in depth at their homes, in the presence of their wives who prepared the family meals.

However, sample sizes were extraordinarily small: an average of only 6% of the total cohort in the seven-day surveys and a mere 2% of the large U.S. cohort, according to Keys' data presented in the first five-year follow-up. These are hardly large enough samples from which to draw meaningful conclusions on any connections between diet and CHD, even though the study was designed to select homogeneous population groups. The U.S. cohort of railroad workers, for example, was far from homogeneous as it included both sedentary clerks and physically active switchmen. Keys himself stated that "dietary estimates for averages of 100 or more men" could be in error by as much as 10%, yet the vast majority of his samples consisted of fewer than 50 men, for which the error would be higher.

If Keys' diet–heart hypothesis were correct, the lack of adequate dietary data could explain why CHD incidence on the Greek island of Corfu was much higher than on the island of Crete, despite the Greeks on Corfu eating less fat; fat consumption on both islands was seasonally lower during the Christian period of Lent. The validity or nonvalidity of the results for apparently healthier Crete is important, because the sparse Cretan data formed the basis of the so-called Mediterranean diet that was promoted in the 1990s.

But whatever the deficiencies of Keys' crowning achievement, the Seven Countries Study has not been the only epidemiological investigation of dietary fat and CHD, although it is the only international one. A handful of regional and national studies, some of which also involved Keys, had been made earlier and dozens more have been conducted since, many of them much larger than the Seven Countries Study. But they don't all agree with the conclusions reached by Keys. During his lifetime, Keys vociferously criticized anyone who dared to challenge his cherished hypothesis. Toward one critic, a Texas academic biochemist who penned a poorly argued but mostly sound critique of the saturated fat hypothesis,[1] Keys was particularly scathing, accusing the attacker of myopia and saying that his article reminded him of "one of the distorting mirrors in the hall of jokes at the county fair."[2] Until recently, many of the early epidemiological investigations of CHD and

---

[1] Reiser, Raymond. 1974. "Saturated Fat in the Diet and Serum Cholesterol Concentration: A Critical Examination of the Literature." *American Journal of Clinical Nutrition*, 26, 524-555.

[2] Keys, Ancel, Francisco Grande and Joseph T. Anderson. 1974. "Bias and Misrepresentation Revisited: Perspective on Saturated Fat." *American Journal of Clinical Nutrition*, 27, 188-212.

dietary fat were simply ignored, despite a massive weight of evidence that conflicted with the diet–heart hypothesis.[1]

One investigation that wasn't ignored was the Framingham Heart Study, begun in 1950 in Framingham, Massachusetts, although it focused on cholesterol levels rather than diet. The goal was to study the incidence of CHD in a group of 5,127 men & women, aged from 30 to 59. The study is most well-known for its finding after six years that the onset of CHD, including CHD death, was apparently tied to the serum cholesterol level, as well as to blood pressure, at least in men from 40 to 59. Men in this age range with highly elevated cholesterol levels had more than a threefold higher risk of contracting CHD or dying from it; for women, the risk factor was more modest.[2] However, the study results reported in the 30-year follow-up were at odds with the six-year data, as we'll see in a later section.

Epidemiology, nevertheless, is limited by its purely observational basis. All that any of these nutritional studies could hope to establish was that dietary fat was somehow linked to CHD, not that it was a cause, and that other "confounding" factors such as heavy smoking or high blood pressure were less likely to be connected with the disease. Delineation of cause and effect requires a randomized controlled trial, which is an experiment used to confirm or eliminate a possible cause of the disease by diminishing its effect, for example by reducing saturated fat intake in a nutritional study. The trial diet is closely controlled, more so at least than in an observational study where participants are simply questioned about their diet. In a randomized controlled trial, the study population is divided randomly into two identical groups, with intervention in only one group and the other group used as a control. Confounding factors are thus evenly distributed between the two groups, so the result of the intervention isn't obscured, as the effect of a particular measured factor might be in an observational study. The *only* difference between the two groups is the intervention.

Both epidemiology and clinical trials use sophisticated statistical techniques to analyze the data collected. The likelihood that a particular factor is a cause of the disease is often quantified by calculating the relative risk for that factor.[3] In nutritional studies, the relative risk is usually in the

---

[1] Teicholz, Nina. 2014. *The Big Fat Surprise: Why Butter, Meat and Cheese Belong in a Healthy Diet.* New York: Simon & Schuster, 53-57.

[2] Kannel, William B., Thomas R. Dawber, Abraham Kagan et al. 1961. "Factors of Risk in the Development of Coronary Heart Disease—Six-Year Follow-up Experience: The Framingham Study." *Annals of Internal Medicine*, 55, 33-50.

[3] In the case of nutrition and CHD, the relative risk or risk ratio is the ratio of the probabilities that high compared to low exposure to a risk factor such as saturated fat intake results in CHD. High and low exposure are often taken to be the top and bottom tertiles, quartiles or quintiles of the fat intake or other risk factor, respectively.

range only from 0.5 to 1.5, which corresponds to a 50% lower to 50% higher risk of contracting the disease under study, respectively. In the classic case of smoking and lung cancer, however, epidemiological studies show that the relative risk of smokers falling victim to the disease can be as high as a gigantic 26[1] — which is high enough to remove any uncertainty whatsoever about causation.

As well as the multiple observational studies of the diet–heart hypothesis, there have been a number of randomized controlled trials to investigate the effect on CHD of replacing saturated fats in the diet with various alternatives. While clinical trials are more costly than observational studies, they're also more powerful in being able to identify likely causes of CHD incidence and mortality.

One of the earliest trials was the Anti-Coronary Club Study, which got under way in 1957 in New York city. The study set out to test the diet–heart hypothesis by placing its volunteer subjects on a "prudent diet" designed to lower serum cholesterol, and thus reduce CHD incidence and deaths.[2] A total of 1,277 men aged between 40 and 59 (the same age range as in the Seven Countries Study) were enrolled, 463 of whom made up the control group that maintained their normal eating habits. The prudent diet substituted corn oil, a vegetable oil that is high in polyunsaturated fats, for butter and other high-saturated-fat foods;[3] the men in the experimental diet group were also required to drink at least two tablespoons of corn oil a day. The diet restricted the percentage of total calories from fat to 33%, with no more than 8% of the total from saturated fats.[4]

But despite its good intentions and its success in decreasing serum cholesterol levels, the Anti-Coronary Club clinical trial produced topsy-turvy results. While individuals on the test diet were able to ward off CHD better than their control group counterparts, eight of them had died from CHD after six years, yet not one of the controls. And 18 members of the diet

---

[1] U.S. Centers for Disease Control and Prevention. 2014. "Smoking and Cancer," http://www.cdc.gov/tobacco/data_statistics/sgr/50th-anniversary/pdfs/fs_smoking_cancer_508.pdf. Accessed 11 July 2018.

[2] Jolliffe, Norman, Seymour H. Rinzler and Morton Archer. 1959. "The Anti-Coronary Club; Including a Discussion of the Effects of a Prudent Diet on the Serum Cholesterol Level of Middle-Aged Men." *American Journal of Clinical Nutrition*, 7, 451-462.

[3] Chemically, saturated fats have no double bonds between the carbon atoms in the fatty acid chain, because all carbon atoms are saturated with hydrogen atoms. Unsaturated fats have one (monounsaturated) or more (polyunsaturated) double bonds. Saturated fats, which include butter and lard, are usually solid at room temperature, whereas unsaturated fats are liquid at room temperature.

[4] Christakis, George, Seymour H. Rinzler, Morton Archer and Arthur Kraus. 1966. "Effect of the Anti-Coronary Club Program on Coronary Heart Disease Risk-Factor Status." *Journal of the American Medical Association*, 198, 597-604.

club had died from other causes during that period, compared with only six in the control group. Even though these numbers are relatively small, the apparent inverse correlation between dietary fat intake and death, either from CHD or for other reasons, was mystifying at the time — though as we'll see, the lack of any correlation at all has become part of a pattern in such clinical studies. Although eating less saturated fat definitely lowers cholesterol levels, it doesn't appear to help us live longer or avoid dying from CHD. The same conclusion was reached in at least two epidemiological studies: at the 25-year follow-up in the Seven Countries Study,[1] and at the 30-year point in the Framingham Study as noted above.

The Anti-Coronary Club Study was followed in the 1960s by several other randomized controlled trials on Keys' hypothesis. What's astounding is that although the Anti-Coronary trial had not shown that the prudent low-fat diet reduced the risk of dying from CHD, and the other trials were seriously flawed, all these studies are frequently cited as evidence in support of the diet–heart hypothesis. The flaws include unstable study populations in which members of the intervention group came and went, drastically changing the group's composition; lack of adequate control of the experimental diet; and significantly different heavy smoking rates between the control and experimental groups.

In an attempt to come up with a more reliable and larger controlled trial, the U.S. National Institutes of Health (NIH) initiated the Multiple Risk Factor Intervention Trial (MRFIT) that ran from 1973 to 1982. The MRFIT enlisted 12,866 men aged between 35 and 57 who had extremely high cholesterol levels,[2] high enough to be considered at imminent risk for a heart attack, and divided them into intervention and control groups of almost equal size. As the trial's title suggests, the men were at risk for more than one reason, so those in the intervention group received multiple interventions. These included treatment for high blood pressure, counseling on cigarette smoking, and dietary advice for cutting back serum cholesterol levels — which consisted primarily of distributing cookbooks from the American Heart Association (AHA), accompanied by a list of supermarket products compatible with the low-fat diet the intervention group were required to

---

[1] Menotti, A., H. Blackburn, D. Kromhout et al. 2001. "Cardiovascular Risk Factors as Determinants of 25-Year All-Cause Mortality in the Seven Countries Study." *European Journal of Epidemiology*, 17, 337-346.
[2] Above 290 mg/dL. Currently in the U.S., a total cholesterol level over 240 mg/dL is considered high.

follow. The control group had no interventions and were allowed to eat as they chose.[1]

Nevertheless, no statistically significant difference in either the CHD or overall death rate was observed between the intervention and control groups after seven years, nor 20 years later. As in the Anti-Coronary Club trial, successfully lowering their cholesterol levels did not extend the lives of the MRFIT intervention participants, any more than quitting smoking and reducing their blood pressure did. Although a recent statistical reanalysis of the original MRFIT trial data claimed that CHD incidence (though not the death rate) was lower in the intervention group, this claim is based on inclusion in the analysis of nonfatal coronary artery surgery and nonfatal congestive heart failure, categories of CHD that aren't normally included in such studies.[2]

So just like many of the epidemiological studies on CHD and dietary fat, the results of a number of major clinical trials were either pushed aside or ignored too. By the early 1980s, Keys' diet–heart hypothesis reigned supreme.

The Government Intervenes

Science is built above all on evidence. It's uncommon — or it used to be — for hunches based on flimsy, incomplete evidence to become accepted as the conventional wisdom in scientific circles, let alone the public at large. But that's exactly what happened with the diet–heart hypothesis formulated and vigorously promoted by Keys, even before his celebrated Seven Countries Study was wrapped up. His critics Yerushalmy and Hilleboe had noted the phenomenon in their 1957 paper:

> But quotation and repetition of the suggestive association soon creates the impression that the relationship is truly valid, and ultimately it acquires status as a supporting link in a chain of presumed proof.[3]

Repetition, as any student of propaganda knows, is a surefire way to embed a concept in the public psyche.

---

[1] MRFIT Research Group. 1982. "Multiple Risk Factor Intervention Trial: Risk Factor Changes and Mortality Results." *Journal of the American Medical Association*, 248, 1465-1477.

[2] Stamler, Jeremiah, James D. Neaton, Jerome D. Cohen et al. 2012. "Multiple Risk Factor Intervention Trial Revisited: A New Perspective Based on Nonfatal and Fatal Composite Endpoints, Coronary and Cardiovascular, During the Trial." *Journal of the American Heart Association*, 2012; 1: e003640, 1-7.

In most clinical trials and epidemiological studies of the diet-heart hypothesis, including the Anti-Coronary Club, Seven Countries and Framingham studies, the term "CHD incidence" covers only nonfatal heart attacks (myocardial infarctions), and bouts of chest pain caused by reduced blood flow to the heart (angina pectoris). With further categories of CHD added in the MRFIT reanalysis, it is difficult to compare the results with those of other studies.

[3] Yerushalmy, J. and H. E. Hilleboe. 1957. "Fat in the Diet and Mortality from Heart Disease; a Methodologic Note." *New York State Journal of Medicine*, 57, 2343-2354.

Keys was nothing if not persistent, though no doubt well intentioned. In 1961, the year in which he was featured on the cover of *Time* magazine, Keys was also appointed to the prestigious and influential Nutrition Committee of the AHA. Four years earlier the Nutrition Committee, out of concern that "people want to know whether they're eating themselves into premature heart disease," had published a report evaluating the evidence for and against the connection between dietary fat and atherosclerosis. Apart from deciding that the current evidence fell short, the committee had slapped Keys on the wrist for "uncompromising stands based on evidence that does not stand up under critical examination."[1] Nonetheless, the same committee with Keys on board published another report in 1961, this time asserting that the evidence did indeed suggest a relationship between CHD and the amount and type of fat consumed, and advising Americans to substitute polyunsaturated for saturated fats in their diet.[2]

Even though its language was cautious, the 1961 AHA report was a watershed in the nutritional community. Not only did it publicly endorse Keys' diet–heart hypothesis, it was also the first official recommendation anywhere in the world to cut back on saturated fat in order to prevent CHD and the associated risk of heart attack and stroke. It was the AHA's seal of approval for the prudent low-fat diet. The report, together with exposure from the *Time* magazine article which was published almost simultaneously, was a triumph for Keys personally and helped make his mark for years to come. Keys' growing fame in turn boosted the fortunes of the AHA, as well as the National Heart Institute (NHI), which was the heart research arm of the NIH and had in 1959 issued a joint report with the AHA titled "A Decade of Progress Against Cardiovascular Disease."

Soon after its founding in 1948, the NHI took over administration of the Framingham Heart Study and also began funding heart research, which included the program directed by Keys at the University of Minnesota. In 1961, the AHA and NHI started to plan the National Diet Heart Study, which was to have been an enormous randomized controlled trial on diet and CHD. The trial would have followed 100,000 middle-aged men in several cities for four or five years, dividing them into 10 different diet groups and supplying them with special fat-modified foods. But this grand plan was abandoned after a feasibility study concluded that the trial would be too costly, and that it would be too difficult to recruit and retain so many participants.[3] Instead,

---

[1] Page, Irvine H., Fredrick J. Stare, A. C. Corcoran et al. 1957. "Atherosclerosis and the Fat Content of the Diet." *Circulation*, 16, 163-178.

[2] Page, Irvine H., Edgar V. Allen, Francis L. Chamberlain et al. 1961. "Dietary Fat and its Relation to Heart Attacks and Strokes." *Circulation*, 23, 133-136.

[3] 1968. "The National Diet-Heart Study Final Report." *Circulation*, 37, I-1 - I-428.

it was left to the NIH to initiate the smaller MRFIT trial discussed earlier. In 1976, the NHI was renamed the National Heart, Lung, and Blood Institute (NHLBI); by 1990 the NHLBI's budget was over $1 billion, most funds going to heart research.

Keys continued his unflagging advocacy of the diet–heart hypothesis throughout the period until his retirement in 1972. He was a member or chairman of numerous committees on nutritional issues, as well as one of the architects of the National Diet Heart Study, along with his colleague Jeremiah Stamler of Northwestern University. Stamler went on to direct the MRFIT trial.

In the 1970s, the U.S. Congress stepped into the picture. The Senate Select Committee on Nutrition and Human Needs, chaired by Senator George McGovern, had been established back in 1968 to deal with what was then regarded as the country's most pressing nutritional problem, namely hunger. But a Senate study group in 1976 ruled that the long-lived select committee had served its purpose and recommended that it now become a much smaller subcommittee. Fighting for his committee's political survival, McGovern shrewdly had it tackle an entirely different problem, or imagined problem — which the committee labeled "the epidemic of killer diseases" — by holding hearings on nutrition and disease, and formulating *Dietary Goals for the United States*, a detailed set of dietary guidelines for the American public.[1]

With much fanfare, McGovern announced the first edition of his *Dietary Goals* by press conference in January 1977, immediately setting off a political firestorm. The U.S. Department of Agriculture (USDA), which had long issued dietary guidelines for the public and at that time recommended eating foods from four basic food groups, was pushing hard for official government guidelines. There was also pressure from consumer-interest groups such as the Center for Science in the Public Interest, which was clamoring for more attention to nutrition. But the NIH was ambivalent, the director of the NHLBI testifying at a Senate hearing in February 1977 on diet and cardiovascular disease that:

> There is no doubt that [serum] cholesterol can be lowered by diet....
> The problem with all of these [clinical] trials is that none of them have
> showed a difference in heart attack or death rate in the treated group.[2]

---

[1] Broad, William J. 1979. "NIH Deals Gingerly with Diet-Disease Link." *Science*, 204, 1175-1178.

[2] U.S. Senate Select Committee on Nutrition and Human Needs. 1977. *Dietary Goals for the United States*. Washington, DC: U.S. Government Printing Office, 2nd Edition, December, XXXII, http://hdl.handle.net/2027/uiug.30112023368936. Accessed 11 July 2018.

Strongly influenced by the earlier AHA report and Keys' promotion of his diet–heart hypothesis, *Dietary Goals* called on Americans to combat killer diseases such as CHD, cancer, diabetes and stroke by boosting their consumption of fruits, vegetables, whole grains, poultry, fish and skim milk; lowering their intake of high-fat foods and substituting polyunsaturated for saturated fats; and cutting their consumption of whole milk, eggs, butterfat and foods high in sugar and salt.

As might be expected, the select committee's proposal triggered an avalanche of criticism from the meat, dairy and egg industries, whose food products had been essential in meeting the existing USDA guidelines but were now effectively blacklisted. Protests, especially from the cattle industry in McGovern's home state of South Dakota, led to concessions in the second edition of *Dietary Goals* published in December 1977: instead of asking consumers to "decrease consumption of meat," the guidelines now recommended that people should simply reduce their intake of animal fat. The new edition also exempted young children from the advice on reducing whole milk and egg consumption.

But the second edition wasn't the last word. Although it was more than enough to satisfy the USDA as a policy document, the agency, unsure of the science behind it, called on the NAS for a scientific evaluation of the *Dietary Goals* guidelines prior to their planned release to the public. However, the NAS, whose Food and Nutrition Board sets the Recommended Dietary Allowances of nutrients every few years, voiced reservations about the need for any guidelines at all. In a stinging rebuke to the USDA, the Board issued a report in 1980 stating that it considered it "scientifically unsound to make single, all-inclusive recommendations to the public regarding intakes of energy, protein, fat, cholesterol, carbohydrate, fiber, and sodium," and that it did not seem "prudent" to recommend raising the dietary ratio of polyunsaturated to saturated fats.[1]

Still bent on pursuing the guidelines, but having gotten wind of NAS thinking even before its report came out, the USDA had decided to rely instead on a task force established by the American Society for Clinical Nutrition (ASCN) to review the evidence linking diet to chronic disease. The nine-scientist task force included members on both sides of the diet-disease debate; one of the co-chairs was Rockefeller University's Edward Ahrens (1915–2000), a pioneer in cholesterol research who often crossed swords with Keys and the AHA over the diet–heart hypothesis. Yet the panel's 1979

---

[1] U.S. National Research Council, Food and Nutrition Board. 1980. *Toward Healthful Diets*. Washington, DC: National Academies Press, https://web.archive.org/web/2002121508451_3/http://www.ulib.org/webRoot/Books/National_Academy_Press_Books/healthful_diets/index.html. Accessed 11 July 2018.

report was no more conclusive than the previous statements made by the NIH or leaked from the NAS. The only clear-cut connections found between diet and chronic disease were those between alcohol and liver disease, and between sugar and tooth decay — hardly startling observations. The evidence for cholesterol or saturated fat playing a significant role in CHD was ambiguous, with considerable disagreement among the panel members.[1]

The USDA, determined to go ahead and issue its guidelines anyway, simply ignored the findings of both the Ahrens report and what it knew was coming from the NAS. In February 1980, *Dietary Guidelines for Americans* was published, the first ever issued to the U.S. public;[2] the UK government proposed similar guidelines three years later.[3] Although preceded in 1961 by the recommendations of the AHA, a professional organization whose sole purpose is to fight heart disease and stroke, *Dietary Guidelines* was the first time that any government had officially endorsed a particular eating pattern. The recommendations not only represented a landmark in government guidance on nutrition, but they also became the foundation of the low-fat diet that was widely adopted around the world for decades. But as just discussed, the guidelines had to be slipped in by the back door, even after thirty years of promotional effort by Keys, and everything that politicians such as McGovern had done to shepherd the new advice through the labyrinth of popular opinion and public policy.

Initially, *Dietary Guidelines* adhered to the recommendations set out in *Dietary Goals*, but without any specific restrictions on the percentages of calories from fat and saturated fats. While both documents linked high consumption of saturated fat to CHD, *Dietary Guidelines* conceded in its section on fat and cholesterol that "There is controversy about what recommendations are appropriate for healthy Americans."[4] *Dietary Guidelines* was the basis for the USDA's 1992 food pyramid, with which most Americans are familiar and which later transformed into MyPlate.

But by 1990, *Dietary Guidelines* included the same limits on fat intake spelled out in the earlier *Dietary Goals*: a maximum of 30% of total calories from all fat, with no more than 10% from saturated fat. The limit on saturated

[1] Ahrens, Edward H. Jr. 1979. "The Evidence Relating Six Dietary Factors to the Nation's Health." *American Journal of Clinical Nutrition*, 32, 2627-2631.
[2] U.S. Department of Agriculture. 1980. *Nutrition and Your Health: Dietary Guidelines for Americans*. Washington, DC: U.S. Government Printing Office, https://www.cnpp.usda.gov/sites/default/files/dietary_guidelines_for_americans/1980thin.pdf. Accessed 11 July 2018.
[3] UK National Advisory Committee on Nutrition Education. 1983. *A Discussion Paper on Proposals for Nutritional Guidelines for Health Education in Britain*. London: Health Education Council.
[4] U.S. Department of Agriculture, 1980, 11-12.

fat is the same as that recommended by the U.S. government today,[1] though the AHA suggests a lower limit for people with high cholesterol.[2] From 2005 to 2015, the recommended limit on total fat consumption was from 20% to 35% of calories; in 2015, however, the limit on total fat abruptly disappeared from both the government and AHA recommendations. *Dietary Goals* noted that average consumption of fat in the U.S. as a percentage of calories had jumped from 32% in 1910 to over 40% in 1960. For comparison, Keys had discovered in 1956 that middle-aged Japanese men living in their native Japan got only 9% to 14% of their calories from fat, a number that shot up to a whopping 39% for Japanese men living in Los Angeles in the U.S.[3]

The NIH, long reluctant to commit itself to supporting the new guidelines, finally came aboard in 1984 when it convened one of its ongoing "consensus development" conferences, on lowering serum cholesterol to prevent heart disease. The conference was the outcome of an NHLBI randomized controlled trial, initiated in the 1970s and intended to answer the question about Keys' diet–heart hypothesis once and for all — whether a diet high in saturated fat, which elevates serum cholesterol levels, increases the risk of contracting CHD. Previous clinical trials to verify the hypothesis, such as the Anti-Coronary Club Study, had been inconclusive as we've noted. However, unlike the Anti-Coronary Club trial or NIH's MRFIT, diet itself wasn't the main emphasis of the NHLBI trial, with a cholesterol-lowering drug treatment being used as the intervention instead. Both the intervention and control groups, nevertheless, were required to follow a cholesterol-reducing diet.[4]

Despite its promise, the NHLBI trial was as disappointing as its predecessors. The drug did reduce serum cholesterol marginally,[5] a finding that marked the beginning of a major NIH thrust in the 1980s to encourage Americans to trim their cholesterol levels, and the later widespread

---

[1] U.S. Department of Health and Human Services and U.S. Department of Agriculture. 2015. "Dietary Guidelines for Americans 2015—2020," 8th Edition, December, http://health.gov/dietaryguidelines/2015/guidelines/. Accessed 11 July 2018.

[2] American Heart Association. 2015. "The Facts on Fats: 50 Years of American Heart Association Dietary Fats Recommendations," June, http://www.heart.org/idc/groups/heart-public/@wcm/@fc/documents/downloadable/ucm_475005.pdf. Accessed 11 July 2018.

[3] Keys, Ancel and Francisco Grande. 1957. "Role of Dietary Fat in Human Nutrition III — Diet and the Epidemiology of Coronary Heart Disease." *American Journal of Public Health and the Nations Health*, 47, 1520—1530.

[4] 1984. "The Lipid Research Clinics Coronary Primary Prevention Trial Results : I. Reduction in Incidence of Coronary Heart Disease." *Journal of the American Medical Association*, 251, 351-364.

[5] The drug used was cholestyramine. Combined with the low-fat diet, this drug reduced total cholesterol levels by an average of 13% in the intervention group, compared to 5% in the controls; a reduction of 5% in both groups was therefore attributed to diet alone.

introduction of cholesterol-lowering statin drugs. But the incidence of CHD was only slightly lower in the trial's intervention group and there was no difference in the overall death rate between the two groups, in common with results of other controlled trials.

Nonetheless, the 1984 NIH consensus conference overstepped its mandate to review the evidence for the diet–heart hypothesis by not only declaring that lowering high cholesterol levels would reduce the risk of heart attacks "beyond a reasonable doubt," but also by making the astounding recommendation that all Americans above the age of two should "adopt a diet" that reduced their consumption of saturated fat and cholesterol, much like that proposed in *Dietary Goals*. This recommendation for the entire population was based primarily on the results of the NHLBI controlled trial. Diet, however, wasn't a major factor in the trial as mentioned above, and the trial was conducted only on middle-aged men, so its extrapolation to women and children was dubious at best. Critics rightly pointed out that, while a low-fat diet might benefit men with very high cholesterol levels, there was no evidence that the diet would be safe for children or the elderly or those with already low cholesterol levels, among which groups there could well be side effects. And along with other clinical trials, the NHLBI trial had failed to show that lowering serum cholesterol saves lives.[1]

Over the years since the consensus conference, the NIH has backed off from its then imperious advocacy of Keys' hypothesis. Nevertheless, its recommendations at the time on diet, coupled with the government's *Dietary Guidelines*, were enough to anoint the low-fat diet as the standard diet worldwide in the 1980s and 1990s. I remember that era myself, when different versions of the diet and low-fat cookbooks proliferated, along with endless magazine and newspaper articles on cutting fat and supposedly living better. That the diet wasn't suitable for all nor particularly healthy because it replaced fat with refined carbohydrates, which contribute to other health problems such as obesity, remained to be discovered.

The largest ever, long-term randomized trial of the low-fat diet was the Women's Health Initiative (WHI), launched across the U.S. in 1993 — 13 years after the USDA had announced its *Dietary Guidelines* and more than 30 years after the AHA had first given its blessing to the prudent low-fat diet. In the WHI trial, the health of 48,835 postmenopausal women aged from 50 to 79 was studied for eight years, to see how dietary intervention would affect several types of cancer including breast and colorectal, in addition to CHD and stroke. As well as CHD, cancer had been linked to fat consumption since the late 1970s. The intervention group, which made

---

[1] Kolata, Gina. 1985. "Heart Panel's Conclusions Questioned." *Science*, 227, 40-41.

up 40% of the participants, received group and individual counseling from trained nutritionists to follow a diet low in fat and high in vegetables, fruits and grains. One of the specific goals of the WHI trial was to reduce total fat intake to a low 20% of calories.[1]

But the trial results were a resounding rejection not only of the efficacy of the low-fat diet, but also of the validity of the diet–heart hypothesis, at least for women. After eight years, the diet was found to have had no effect on either CHD incidence or deaths, nor on any form of cancer except possibly ovarian.[2] About the only success the trial could claim in terms of its goals was that the intervention group had reduced their average total fat consumption to 29% of calories from an initial baseline of 37%, and their saturated fat from 12% to 9%, after six years. But this achievement hardly vindicates the low-fat diet in terms of health benefits, any more than the observation by WHI investigators that women who initially ate the most fat and adhered to the diet showed the lowest rates of breast cancer at the trial's end; the breast cancer rate was unchanged in those who started out eating the least fat. Overall, the results were a source of consternation to both the WHI investigators and advocates of the low-fat diet.

Equally puzzling findings resulted from a short-term study of the low-fat diet conducted in the 1990s on several hundred U.S. Boeing employees, both men and women. Like many other observational studies and clinical trials before it, the Boeing study examined the connection between diet and cholesterol levels. The diet in this case limited total fat intake to 30% of calories, with a maximum of a very low 7% from saturated fat.

Although the diet successfully lowered cholesterol levels, just as the prudent diet had done in the Anti-Coronary Club Study more than 30 years earlier, the Boeing results were as startling as the Anti-Coronary trial's lack of evidence for the diet–heart hypothesis. First, the Boeing study discovered that the low-fat diet no longer reduced serum cholesterol if fat intake fell below about 25% of calories.[3] Second, while "bad" cholesterol levels went down significantly on the diet, so did "good" cholesterol in women, though

[1] Howard, Barbara V., Linda Van Horn, Judith Hsia et al. 2006. "Low-Fat Dietary Pattern and Risk of Cardiovascular Disease: The Women's Health Initiative Randomized Controlled Dietary Modification Trial." *Journal of the American Medical Association*, 295, 655-666.

[2] Prentice, Ross L., Cynthia A. Thomson, Bette Caan et al. 2007. "Low-Fat Dietary Pattern and Cancer Incidence in the Women's Health Initiative Dietary Modification Randomized Controlled Trial." *Journal of the National Cancer Institute*, 99, 1534-1543.

[3] Knopp, Robert H., Barbara Retzlaff, Carolyn Walden et al. 2000. "One-Year Effects of Increasingly Fat-Restricted, Carbohydrate-Enriched Diets on Lipoprotein Levels in Free-Living Subjects." *Proceedings of the Society for Experimental Biology and Medicine*, 225, 191-199.

not men.[1] What this means is that women who had followed the low-fat diet had apparently increased their risk for CHD, a disturbing finding that was brushed aside by low-fat diet proponents.

If you're getting the impression that the much ballyhooed low-fat diet wasn't very effective, you're right. Bad news for the diet continued to accumulate. A United Nations (UN) review of available data in 2008 found that several randomized controlled trials of the diet, including the WHI trial, "have not found evidence for beneficial effects of low-fat diets," and remarked that there wasn't any convincing evidence either for significant connections between dietary fat and CHD or cancers.[2]

As a consequence of these failures, the low-fat diet was edged out of the spotlight in the 1990s by the Mediterranean diet. The Mediterranean diet was heavily promoted not by governments but by commercial interests, primarily the olive oil industry. The diet itself harks back to Keys and his fascination during the Seven Countries Study with Crete, where he found CHD was relatively rare and life spans were long, yet the populace consumed copious quantities of high-fat olive oil. However, olive oil is low in saturated fat — the fat demonized by Keys that became the bane of the low-fat diet — and high in monounsaturated fat, considered healthier than polyunsaturated fat. The Mediterranean diet emphasizes olive oil, along with vegetables, fruits, legumes and whole grains; moderate amounts of seafood, poultry, nuts and dairy products; and limited servings of lean red meat. Keys himself practiced what he preached, building a retirement villa on the Italian coast and adopting both the Mediterranean diet and a Mediterranean lifestyle until he died.

Randomized trials of the popular Mediterranean diet have demonstrated its superiority to the officially sanctioned low-fat diet in lowering serum cholesterol,[3] though a large Spanish trial showed no clear benefit in reducing the risk of CHD incidence or death,[4] just like the low-fat diet findings. On the other hand, a few small randomized trials suggest that the Mediterranean

[1] Walden, Carolyn E., Barbara M. Retzlaff, Brenda L. Buck et al. 1997. "Lipoprotein Lipid Response to the National Cholesterol Education Program Step II Diet by Hypercholesterolemic and Combined Hyperlipidemic Women and Men." *Arteriosclerosis, Thrombosis, and Vascular Biology*, 17, 375-382.

[2] UN Food and Agriculture Organization. 2010. *Fats and Fatty Acids in Human Nutrition: Report of an Expert Consultation, 10-14 November 2008, Geneva*, Food and Nutrition Paper 91. Rome: Food and Agriculture Organization of the United Nations, 2, 13.

[3] Shai, Iris, Dan Schwarzfuchs, Yaakov Henkin et al. 2008. "Weight Loss with a Low-Carbohydrate, Mediterranean, or Low-Fat Diet." *New England Journal of Medicine*, 359, 229-241.

[4] Estruch, Ramón, Emilio Ros, Jordi Salas-Salvadó et al. 2013. "Primary Prevention of Cardiovascular Disease with a Mediterranean Diet." *New England Journal of Medicine*, 368, 1279-90.

diet can lower the incidence and growth rate of certain types of cancer, such as breast and prostate cancer. However, what's amazing about the Mediterranean diet is how quickly it established such a strong foothold in numerous countries, after originating with a mere 63 men on Crete. Keys was an unequaled driving force in the nutritional world.

A New Look at Old Data

Altogether there have been more than a hundred epidemiological studies and a smaller number of randomized controlled trials of the diet–heart hypothesis, or of specific diets designed to cut back on the saturated fat denounced by the hypothesis. While we have looked at only a few of the many studies and trials in this chapter, it will be apparent that evidence for the diet–heart hypothesis has often been inconclusive or even conflicting. In an effort to shed light on the problem, nutritional and medical researchers have recently resorted to new statistical methods of combining data from individual studies or trials. The hope was to obtain more precise estimates, from the pooled evidence, of the relative risk for CHD or of the effect of a particular diet.

The combined analysis is usually called a meta-analysis in the case of observational studies, and a systematic review in the case of clinical trials.[1] Because the number of participants in specific studies or trials and thus their accuracy varies widely, each study or trial in the combined analysis is weighted according to its estimated accuracy.[2] And statistical adjustments are made to account for differences from study to study in the observation period, the age range of the participants, consumption of different nutrients and confounding factors such as smoking.

Two important conclusions can be drawn from the various meta-analyses and systematic reviews conducted to date. The first and indisputable finding is that eating saturated fat doesn't shorten our lives: there's no convincing evidence for any connection between saturated fat intake and *deaths* from CHD, or indeed deaths from any cause at all. This finding simply confirms what earlier individual trials such as the MRFIT and WHI trial had deduced, and is a very big nail in the coffin of Keys' diet–heart hypothesis. Yet, the science aside, the U.S. government and the AHA continue to advocate restricting consumption of saturated fat.

The second conclusion is that there appears to be a small 10% to 15% reduction in the risk for CHD *events*, not deaths, when saturated fat in the diet is replaced by polyunsaturated fats. By CHD events are usually meant

---

[1] Haidich, A. B. 2010. "Meta-Analysis in Medical Research." *Hippokratia*, 14, 29-37, http://www.ncbi.nlm.nih.gov/pmc/articles/PMC3049418/. Accessed 11 July 2018.

[2] Typically, individual studies or trials are weighted inversely by their statistical variance; the variance decreases as the number of participants increases.

nonfatal heart attacks and bouts of angina pain; some studies, however, include CHD deaths in the events category, which is confusing. But just as the results of the many individual studies and trials of the diet–heart hypothesis are all over the lot, conflicting meta-analyses and systematic reviews only add to the mystery, despite the fact that the various analyses and reviews draw on many of the same studies and trials.

For example, while two meta-analyses, each including more than 150,000 participants,[1, 2] have shown no association between saturated fat intake and CHD events, three other meta-analyses have found that the risk for CHD events is up to 15% lower when saturated fat in the diet is replaced by a polyunsaturated fat.[3, 4, 5] A 2015 clinical trial review, conducted by the prestigious Cochrane Collaboration, found a 13% lower risk for CHD events when saturated fat intake was reduced but not replaced by another nutrient.[6] On the other hand, a 2016 review by a different group found no evidence for any reduced risk of CHD events when the polyunsaturated fat linoleic acid is substituted for saturated fat in the diet; the review also suggested that the risk of CHD death for those over 65 may actually go up if such a substitution is made.[7] The 2015 review included the very large WHI trial that showed no effect of a low-fat diet on CHD incidence, while the 2016 review that found essentially the same result as the WHI trial excluded it.

The 2016 review, which completely refutes the diet–heart hypothesis, included previously unpublished data from two randomized trials that also failed to support the hypothesis — the Sydney Diet Heart Study and the

[1] Siri-Tarino, Patty W., Qi Sun, Frank B. Hu and Ronald M Krauss. 2010. "Meta-Analysis of Prospective Cohort Studies Evaluating the Association of Saturated Fat with Cardiovascular Disease." *American Journal of Clinical Nutrition*, 91, 535—546.

[2] de Souza, Russell J., Andrew Mente, Adriana Maroleanu et al. 2015. "Intake of Saturated and Trans Unsaturated Fatty Acids and Risk of All Cause Mortality, Cardiovascular Disease, and Type 2 Diabetes: Systematic Review and Meta-Analysis of Observational Studies." *BMJ*, 2015; 351: h3978, 1-16.

[3] Skeaff, C. Murray and Jody Miller. 2009. "Dietary Fat and Coronary Heart Disease: Summary of Evidence from Prospective Cohort and Randomised Controlled Trials." *Annals of Nutrition and Metabolism*, 55, 173-201.

[4] Farvid, Maryam S., Ming Ding, An Pan et al. 2014. "Dietary Linoleic Acid and Risk of Coronary Heart Disease: A Systematic Review and Meta-Analysis of Prospective Cohort Studies." *Circulation*, 130, 1568-1578.

[5] Li, Yanping, Adela Hruby, Adam M. Bernstein et al. 2015. "Saturated Fats Compared with Unsaturated Fats and Sources of Carbohydrates in Relation to Risk of Coronary Heart Disease: A Prospective Cohort Study." *Journal of the American College of Cardiology*, 66, 1538-1548.

[6] Hooper, Lee, Nicole Martin and Asmaa Abdelhamid. 2015. "Cochrane Corner: What are the Effects of Reducing Saturated Fat Intake on Cardiovascular Disease and Mortality?" *Heart*, 101, 1938-1940.

[7] Ramsden, Christopher E., Daisy Zamora, Sharon Majchrzak-Hong et al. 2016. "Re-Evaluation of the Traditional Diet-Heart Hypothesis: Analysis of Recovered Data from Minnesota Coronary Experiment (1968-73)." *BMJ*, 2016; 353: i1246, 1-17.

Minnesota Coronary Experiment (MCE), which began in 1966 and 1968, respectively, both ending in 1973. The 2015 Cochrane review, on the other hand, included both trials but not the formerly unpublished data. It's ironic that the co-principal investigator of the MCE trial was Keys himself. That critical analyses and other raw data from the MCE trial languished for more than 30 years before being recovered, and that even preliminary results of the trial weren't published until 1989,[1] smacks strongly of deliberate data suppression. As we've already seen, this wasn't the first time that evidence conflicting with the diet–heart hypothesis had been ignored.

The authors of the 2016 review even speculated whether complete publication of the Sydney and Minnesota results at an earlier stage might have altered the government policy decisions that led to the *Dietary Guidelines* discussed in the previous section. They went on to call our current understanding of nutrition "rudimentary" and, in a thinly veiled swipe at diet–heart proponents, suggested that we could be more humble in confronting the lack of unambiguous evidence for the diet–heart hypothesis.

The lack of evidence for a connection between saturated fat consumption and deaths from any cause at all reveals itself not only in meta-analyses and systematic reviews, but also in the long-term results of several of the older observational studies. As we know, eating more saturated fat raises cholesterol levels, while eating less saturated fat lowers them. The six-year follow-up in the Framingham Heart Study, discussed earlier in the chapter, linked CHD incidence and deaths in men aged between 40 and 59 to elevation of their serum cholesterol levels, as well as elevated blood pressure. But the 30-year follow-up to the Framingham Study found exactly the reverse. For both men and women over the age of 50, there was no increase in deaths from all causes, including CHD, with rising cholesterol; in fact, a puzzling association was noted in this age group between mortality and *falling* cholesterol levels.[2]

Likewise, the 25-year follow-up in the Seven Countries Study, published in 2001 long after Keys had retired, found no significant relationship between cholesterol and the overall death rate, just as had been observed in the 15-year follow-up that was coauthored by Keys. Age, blood pressure and smoking habits were shown to be much more reliable indicators of mortality

[1] Frantz, Ivan D. Jr, Emily A. Dawson, Patricia L. Ashman et al. 1989. "Test of Effect of Lipid Lowering by Diet on Cardiovascular Risk. The Minnesota Coronary Survey." *Arteriosclerosis, Thrombosis, and Vascular Biology*, 9, 129-135.
[2] Anderson, Keaven M., William P. Castelli and Daniel Levy. 1987. "Cholesterol and Mortality: : 30 Years of Follow-Up from the Framingham Study." *Journal of the American Medical Association*, 257, 2176-2180.

than cholesterol. And at both 15 and 25 years, fewer than 30% of all deaths in the study were a result of CHD.

Although the authors of the 25-year follow-up asserted that controlling one's blood pressure, smoking and cholesterol could "influence positively life expectancy and reduce death rates," a 1991 study based on a computer model had previously estimated that a 35-year-old American man would gain just eight months of life by reducing his serum cholesterol to a low level.[1] The same year, another study led by epidemiologist Warren Browner had concluded that the increase in life expectancy for all Americans would be only three to four months on average if they restricted their total fat intake over a lifetime to 30% of calories,[2] the same limit recommended in the U.S. government's 1977 *Dietary Goals*. As modest and as insignificant to many people as this estimate was, it nevertheless generated an enormous backlash from Browner's government funding agency which tried unsuccessfully to stop publication of his research.[3] The study's predicted lack of benefit from the low-fat diet that the government was pushing at the time was surely seen as sabotage.

The failure to verify Keys' diet–heart hypothesis after more than 60 years of effort brings us full circle. Back in the 1960s, not long after Keys proposed his hypothesis, George Mann (1917–2013), who was a medical researcher at Vanderbilt University and associate director of the Framingham Study, led a team to Tanzania to study CHD in the tribal Maasai people. Mann knew that they ate nothing but milk, meat and cow blood — a diet consisting almost entirely of animal products that provided an enormous 66% of calories from fat, with much of this being saturated (as well as monounsaturated) fat. If Keys were right, the Maasai should have been plagued by heart disease. But Mann found just the opposite: Maasai men had extremely low cholesterol levels and his autopsies showed little evidence for atherosclerosis, contradicting the diet–heart hypothesis.[4] However, Mann's results don't imply that excessively high fat consumption is necessarily healthy, except

---

[1] Tsevat, Joel, Milton C. Weinstein, Lawrence W. Williams et al. 1991. "Expected Gains in Life Expectancy from Various Coronary Heart Disease Risk Factor Modifications." *Circulation*, 83, 1194-1201. By low was meant under 200 mg/dL.

[2] Browner, Warren S., Janice Westenhouse and Jeffrey A. Tice. 1991. "What if Americans Ate Less Fat? A Quantitative Estimate of the Effect on Mortality." *Journal of the American Medical Association*, 265, 3285-3291.

[3] Taubes, Gary. 2001. "What if Americans Ate Less Saturated Fat?" *Science*, 291, 2538.

[4] Mann, G. V., R. D. Shaffer, R.S. Anderson et al. 1964. "Cardiovascular Disease in the Masai" (Masai is an older British spelling of the more correct Maasai). *Atherosclerosis*, 4, 289-312.

for the Maasai. Subsequent studies showed that in this case, genetic factors are probably important.[1, 2]

Bias And Politics Invade Science

Unlike continental drift and evolution, the more evidence that was collected on the diet–heart hypothesis, the murkier the picture became. The more epidemiological studies and randomized controlled trials that were carried out, the less clear it was that any form of dietary intervention could ward off heart attacks or prolong life spans. But despite the flaws in some of the earlier studies and trials, the majority of the investigations overall represent sound science, conducted by well-intentioned scientists following the scientific method. Science only began to come off the rails when two external forces intruded: confirmation bias, and political interference. Both of these led to abuse and corruption of the nutritional science world.

Confirmation bias describes the all too human tendency to seek out information that confirms one's preconceptions or hypotheses, and to cast aside or ignore any contradictory evidence. In the case of dietary fat, the bias began with Keys and his tunnel vision on the diet–heart hypothesis.

As discussed earlier, Keys' intense focus on the association of CHD with saturated animal fat in the diet was criticized by Yerushalmy and Hilleboe, who not only demonstrated that the correlation was much weaker than Keys claimed, but also found that the correlation between CHD and animal protein was just as strong. At the annual meeting of the American Public Health Association in 1956, the summary of a symposium on dietary fat in nutrition concluded that "The value of a diet restricted in fat in the prevention of atherosclerosis in Americans has not been demonstrated." This summary also remarked that drastically cutting fat intake from 40% to only 10% of calories by Americans hadn't been found to lower their serum cholesterol to the levels observed in the Japanese and Nigerians, whose diets have a paucity of fat.[3] But that didn't hold Keys back from declaring the following year that "Experimental, theoretical, and epidemiologic evidence implicates the diet, and especially the fats in the diet" in differences in CHD from country to country — or from pushing ahead with his Seven Countries Study, with its highly selective choice of countries and population groups.

[1] Taylor, C. Bruce and Kang-Jey Ho. 1971. "Studies on the Masai." *American Journal of Clinical Nutrition*, 24, 1291-1293.
[2] Wagh, Kshitij, Aatish Bhatia, Gabriela Alexe et al. 2012. "Lactase Persistence and Lipid Pathway Selection in the Maasai." *PLoS One*, 7(9): e44751, 1-12.
[3] Olson, Robert E. 1957. "Role of Dietary Fat in Human Nutrition: V. Summary." *American Journal of Public Health and the Nations Health*, 47, 1537-1541.

Keys had little tolerance for those who disagreed with him and vigorously defended his diet–heart hypothesis against criticism. While some of this behavior reflects his forceful personality, rejection of contrary evidence is also part of confirmation bias: Keys sought out evidence that conflicted with his hypothesis and went out of his way to demolish it. His sarcastic panning of the Texas biochemist's critique on saturated fat ran to 25 pages, which would be long even for an original article; he was granted a three-page response to an observational study on heart attack deaths in a Pennsylvania town,[1] when the study report itself took up only five pages in the same journal; and he wrote a 10-page paper disparaging the "mountain of nonsense" from a British physiologist who had proposed that sugar, not saturated fat, was the culprit in CHD,[2] a notion that has recently been revived.

However, Keys' strong opinions on their own weren't enough to change the course of nutrition science. The diet–heart hypothesis only became nutritional dogma when his bias was institutionalized, primarily through his influence at the AHA. Founded in 1924, the AHA grew rapidly after the organization was selected to receive the proceeds of almost $2 million from the "Truth or Consequences" radio contest in 1948. By the time that Keys was appointed to the Nutrition Committee in 1961, the AHA was the preeminent U.S. authority on heart disease and a major lobbying and political force in public health. By endorsing the diet–heart hypothesis and advocating the low-fat diet, the AHA helped perpetuate Keys' biased view of saturated fat. Only now, some 50 years later, is the whole paradigm being questioned.

Apart from dietary advice, the AHA and the closely linked NHI, later the NHLBI, propagated Keys' bias by fostering what Vanderbilt's Mann called a "heart Mafia." This nutritional research elite dominated the AHA and NHI and, Mann maintained, usurped the available funding for heart research through a peer-review system that rewarded conformity and excluded those like Mann who disputed the diet–heart consensus.[3] Although Mann was always a controversial figure, others such as Rockefeller's Ahrens experienced the very same difficulty in obtaining research grants, solely because of their opposition to the NHLBI-AHA establishment.

Confirmation bias toward affirming the diet–heart hypothesis has endured to the present day, with the belief persisting that too much saturated fat is bad for us all, whatever our cholesterol levels — although

---

[1] Keys, Ancel. 1966. "Arteriosclerotic Heart Disease in Roseto, Pennsylvania." *Journal of the American Medical Association*, 195, 93-95.

[2] Keys, A. 1971. "Sucrose in the Diet and Coronary Heart Disease." *Atherosclerosis*, 14, 193-202.

[3] Mann, George V. 1978. "Coronary Heart Disease — The Doctor's Dilemma." *American Heart Journal*, 96, 569-571.

Dietary Guidelines in 2015 finally conceded that restricting total fat intake is unnecessary, as noted before. The bias even extends to the recent meta-analyses and systematic reviews mentioned in the previous section, the purpose of which was to clarify the big picture on CHD and eating habits, not to reinforce any particular view. However, what stands out is that neither of the two meta-analyses upholding the diet–heart hypothesis have attracted any scientific comments in the medical literature (with the exception of a letter calling the results of that particular meta-analysis "fragile"[1]), while the meta-analyses and trial reviews showing *no* link between CHD events and saturated fat have all drawn highly critical, even dismissive comments.

One of the longest and most critical responses was from MRFIT director Stamler, a member of the NHLBI-AHA inner circle of Keys devotees, who described the analysis in question as "problematic in its thrust."[2] Other critical comments came from Harvard epidemiologist Walter Willett,[3] champion of the Mediterranean diet and instrumental in having trans fats banned, though he overstated the scientific case; and from low-fat diet guru Dean Ornish.[4] Such denouncement of contradictory evidence is unscientific and precisely the tactic that Keys himself employed in the 1960s and 1970s to shore up his diet–heart hypothesis.

The biased campaign to silence critics of Keys' hypothesis has recently spread to the feature pages of the prominent medical journal *BMJ* (formerly *British Medical Journal*). The September 23, 2015 issue carried an article by investigative journalist and author Nina Teicholz questioning the science behind the then soon-to-be-released U.S. *Dietary Guidelines*,[5] which are reissued every five years. The article, highly critical of the guidelines advisory committee whose scientific standards she called weak and subject to bias, was based on research Teicholz had done for her 2014 best-selling book, *The Big Fat Surprise: Why Butter, Meat and Cheese Belong in a Healthy Diet.* The response

[1] Hoenselaar, Robert. 2015. "Letter by Hoenselaar Regarding Article, 'Dietary Linoleic Acid and Risk of Coronary Heart Disease: A Systematic Review and Meta-Analysis of Prospective Cohort Studies'." *Circulation*, 2015; 132: e20.

[2] Stamler, Jeremiah. 2010. "Diet-Heart: A Problematic Revisit." *American Journal of Clinical Nutrition*, 91, 497-499.

[3] Willett, Walter C. 2016. "Re: Re-Evaluation of the Traditional Diet-Heart Hypothesis: Analysis of Recovered Data from Minnesota Coronary Experiment (1968-73)," Rapid Response, 16 April. *BMJ*, 2016; 353: i1246.

[4] Ornish, Dean. 2015. "Re: Intake of Saturated and Trans Unsaturated Fatty Acids and Risk of All Cause Mortality, Cardiovascular Disease, and Type 2 Diabetes: Systematic Review and Meta-Analysis of Observational Studies," Rapid Response, 12 August. *BMJ*, 2015; 351: h3978.

[5] Teicholz, Nina. 2015. "The Scientific Report Guiding the US Dietary Guidelines: Is it Scientific?" *BMJ*, 2015; 351: h4962, 1-6; Correction for "The Scientific Report Guiding the US Dietary Guidelines: Is it Scientific?" *BMJ*, 2015; 351: h5686.

of the nutrition establishment was fast and ferocious. Just over a month later, a letter signed by 173 doctors, academics and nutritionists appeared in the *BMJ*, claiming that Teicholz's article contained "numerous errors and misrepresentations" and strongly urging the journal to retract it.[1]

Such a response isn't science, any more than the bias that led up to it. Retraction of a journal article is usually reserved for cases of scientific fraud; apart from one error that was corrected a month later, all the so-called errors in the *BMJ* article were either trivial or not errors at all, but rather statements conflicting with the committee's biased interpretation of the meta-analyses and systematic reviews that we just examined. The letter's signatories tellingly included all 14 members of the *Dietary Guidelines* advisory committee, as well as Henry Blackburn, Stamler and Willett, who all belong to the cadre of Keys believers. But it's a credit to the *BMJ* and a ray of hope for science that the journal in 2016 announced that it had found no grounds for retraction of the article, saying that Teicholz's criticisms of the committee were justified. The *BMJ* editor-in-chief added that stronger, unbiased science is urgently needed in the nutritional field.[2]

Bias isn't the only reason that nutrition science lost its way over the dietary fat issue, however. Politics have also played a major role, both internally in the nutrition science community and externally in government and commerce. The internal politics began when the AHA became entangled with nutrition, especially after the AHA Nutrition Committee was created in 1954. Uncertain of its role initially, the committee had by 1979 gained in stature to the point where it was able to declare:

> The rationale is taken that some degree of hypercholesteremia is almost universal in the U.S. population. Thus, the whole population can be considered to be at risk compared to populations with much lower cholesterol levels. This premise can be accepted as a reasonable one on which to recommend universal dietary alteration.[3]

This was an early thumbs-up for the 1980 publication of *Dietary Guidelines*, marking the U.S. government's formal entry into the diet–heart debate, although political involvement at the government level had its origins in Senator McGovern's select committee and its hearings on nutrition and disease three years before.

[1] Liebman, Bonnie. 2015. "Re: The Scientific Report Guiding the US Dietary Guidelines: Is it Scientific?," Rapid Response, 17 December. *BMJ*, 2015; 351: h4962.

[2] Godlee, Fiona. 2016. "Outcome of Post-Publication Review of Article by Nina Teicholz," Rapid Response, 1 December. *BMJ*, 2015; 351: h4962.

[3] Blackburn, Henry. 2005. "American Heart Association Nutrition Committee History," http://www.epi.umn.edu/cvdepi/essay/american-heart-association-nutrition-committee-history/. Accessed 11 July 2018.

The select committee's switch of emphasis from feeding the hungry to nutritional health was driven by the committee's need for political survival. Several powerful senators were committee members, including former Vice President Hubert Humphrey, past presidential nominee George McGovern, future nominee Bob Dole and "Lion of the Senate" Ted Kennedy — all with reputations to protect. The committee was therefore more than ready to take a stand on the growing public concern about diet and diseases such as CHD and cancer. In his January 1977 press conference announcing the first edition of *Dietary Goals*, McGovern pointed out that six of the ten leading causes of death in the U.S. at that time had been linked to diet, and warned that dietary changes over the previous 50 years represented "as great a threat to public health as smoking." Indeed, the committee's hope was that its report would do as much for diet and chronic disease as the 1964 Surgeon General's report had done for cigarettes and lung cancer.

Science at root involves evidence and reason. But there's little room for reason in modern politics, where visceral drives and power plays often dominate logical thought, despite the lip service politicians pay to rational analysis of issues under discussion. The invasion of science by politics is clearly exemplified by the USDA's single-mindedness in pushing ahead with the publication of *Dietary Guidelines*, despite the McGovern committee's acknowledgement in the second edition of *Dietary Goals* that the scientific evidence linking CHD to diet was weak, and despite serious reservations expressed by the NAS Food and Nutrition Board and by Ahrens' ASCN task force. The government's USDA simply rode roughshod over all opposition, responding more to the political mood of the day than to any reasoned assessment of the evidence. Science is poorly served by the political machine.

Science is ill served also by bureaucracies, be they large professional organizations such as the AHA or government agencies such as the NHLBI or the USDA. Bureaucracies by their very nature are cumbersome and inflexible. Once an idea such as the diet–heart hypothesis becomes entrenched in the thinking of a bureaucratic system, it takes on a life of its own and can be very difficult to dislodge. It's hardly surprising that funding for those who disputed the hypothesis all but dried up following the government's intrusion into nutrition science in the late 1970s, aided as it was by bias at the AHA and NHLBI.

Another political influence on nutrition science is of course the food industry. Because nutrition studies and trials are expensive, and funding sources are limited, nutritional researchers have long turned to food companies for financial support, opening the door for outside influence on research and even for corruption. We've already seen how cattle producers lobbied to have a recommendation on eating less meat removed from *Dietary Goals*. On the other hand, frequent criticism of Keys by the meat and dairy industries didn't shake his belief in the diet–heart hypothesis. Other nutrition researchers, however, have been pressured on more than one occasion to deliver results favorable to their financial sponsor. Keys' colleague Blackburn decried the corrupting influence of commercial funding for epidemiology and CHD research, at a time when government funding was shrinking, in a 1992 lecture: "Just as the growing dependence of members of Congress on special interest funds tends to corrupt government, so, too, it corrupts the scientific process."[1]

Yet, the politics aside, the scientific evidence suggests that the causes of heart disease may well lie somewhere else than in dietary fat. Just as low rates of CHD in the fat-loving Maasai may be attributable to genetics, genetics may go a long way toward explaining different rates of CHD from country to country. Perhaps Keys and his believers were barking up the wrong tree from the beginning.

It's thought that genes play a large role in determining body shape and size,[2] so it's quite likely that metabolism is governed in part by our genes too. What this means for diet is that one size definitely does not fit all, certainly not both men and women — a notion that is slowly gaining acceptance in nutrition science, in spite of the hundreds of diets promoted as panaceas for everyone's health. The notion is aptly summarized in the 17th-century nursery rhyme:

> Jack Sprat could eat no fat.
> His wife could eat no lean.
> And so between them both, you see,
> They licked the platter clean.[3]

Even if there are only a handful of different metabolic types, it's a misuse of science for the government to tell us we all need to follow the same diet if we want to stave off heart disease.

---

[1] Blackburn, Henry. 1992. "Ancel Keys Lecture. The Three Beauties: Bench, Clinical, and Population Research." *Circulation*, 86, 1323-1331.

[2] Renki, Margaret. 2010. "Destined to Inherit Your Mom's Body?" *Women's Health on NBCNews.com*, 8 February, http://www.nbcnews.com/id/35254750/ns/health-womens_health/t/destined-inherit-your-moms-body/#.V5PhX_Pn-po. Accessed 11 July 2018.

[3] Opie, Iona and Peter (eds.). 1997. *The Oxford Dictionary of Nursery Rhymes*. Oxford: Oxford University Press, 238. Originally, this nursery rhyme was an allegorical reference to British history.

Chapter 5. Climate Change: Environmental Politics

There's probably no scientific topic more frequently in the news than global warming, often called climate change. A highly controversial subject, it has polarized both scientific and public opinion: on the one hand are those who believe that global warming is largely the result of human activity, and that the science is settled; on the other hand are skeptics who question how much of the warming comes from humans, and how much is the result of natural climate variability. What makes it difficult to discuss the science is that, almost from the beginning in the 1980s, the science has been intertwined with politics — both environmental and governmental. Because of this, the logic and evidence that are the hallmarks of genuine science are often absent from the global warming debate.

As politics have played such a major role, we'll review the politics first in this chapter before taking a look at the science of global warming. We'll see that the brouhaha about catastrophic consequences of man-made warming is based entirely on artificial computer models, and that the actual empirical evidence for a substantial human contribution to warming is flimsy.

The Global Warming Story

The global warming story has its roots in the opposite phenomenon, namely global cooling. From the time that modern warming began around 1850, denoting the end of a prolonged cool period known as the Little Ice Age, the average global temperature has gone up, though in spurts. After climbing by about 0.5° Celsius (0.9° Fahrenheit) from 1910 to 1940, which was part of a return to the warmth before the Little Ice Age, the rising mercury suddenly went into reverse for the next 30 years. The resulting temperature drop, though less than 0.2° Celsius (0.4° Fahrenheit), set off panic in the early 1970s among the media and some scientists, who became convinced that the next regular ice age was upon us.

Although the panic subsided later in the decade when temperatures started to increase again, the episode planted the idea among the media and public that human activities could influence Earth's climate. In the case of the cooling scare, one of the explanations proposed for the temperature decline from 1940 to 1970 was partial blocking of the sun's heat energy by man-made dust in the atmosphere from smog and industrial smokestacks. This effect still exists today, but is much smaller than it was then thanks to clean air acts around the world that have lowered global aerosol emissions. But the temperature rise that commenced in the 1970s shifted attention to potential warming from the ever-mounting presence of the greenhouse gas carbon dioxide (CO_2) in the atmosphere. Much of the extra CO_2 comes from the burning of fossil fuels such as coal, oil and natural gas.

The greenhouse effect, named (though incorrectly) for the process that ripens tomatoes in a glass hothouse, is a well-understood scientific phenomenon. A simplified explanation is that greenhouse gases in the atmosphere act as a radiative blanket around the earth, trapping some of the sun's heat that would otherwise be radiated away.[1] A possible greenhouse connection between increased CO_2 levels and higher temperatures was first postulated by the Swedish chemist Svante Arrhenius (1859–1927).

The notion that human emissions of CO_2 could result in global warming, also put forward by Arrhenius, gained credence in the 20th century. By the time that warming resumed after the mid-century cooling, climate scientists had begun to make highly accurate measurements of the atmospheric CO_2 level, and had developed simple computer models to estimate how much warming might occur as CO_2 built up. I'll discuss the models briefly in the next section, but a 1979 NAS report, based on these models, predicted a hefty increase in global temperatures of between 2° Celsius (3.6° Fahrenheit) and 3.5° Celsius (6.3° Fahrenheit) once the CO_2 level doubles from where it stood in 1850[2] — which is expected to occur later this century. Current

---

[1] Solar radiation is absorbed and radiated by the earth and its atmosphere in two different wavelength regions: absorption takes place at short (ultraviolet and visible) wavelengths, while heat is radiated away at long (infrared) wavelengths. Greenhouse gases in the atmosphere allow most of the incoming shortwave radiation to pass through, but absorb a substantial portion of the outgoing longwave radiation.

[2] U.S. National Research Council, Climate Research Board. 1979. *Carbon Dioxide and Climate: A Scientific Assessment*, Report of an Ad Hoc Study Group on Carbon Dioxide and Climate. Washington, DC: National Academies Press, http://www.nap.edu/catalog/12181.html. Accessed 12 July 2018.

model estimates of the future temperature rise aren't that different, despite massive increases in computing power since 1979. The NAS report, also known as the Charney Report, predicted as well that the warming would lead to noticeable climatic changes.

The possibility of human-induced climate change quickly attracted the notice of politicians in the industrialized world, who organized a series of European meetings on climate starting in 1979, accompanied by U.S. Congressional hearings in the 1980s. At the 1986 hearing of a U.S. Senate Subcommittee on Environmental Pollution, the chairman expressed concern that global warming from greenhouse gases "could, within the next 50 to 75 years, produce enormous changes in a climate that has remained fairly stable for thousands of years."[1]

One climate scientist who was especially vocal about the likely impact on climate of an escalating CO_2 level was James Hansen, an astrophysicist who was head of NASA's Goddard Institute for Space Studies (GISS) from 1981 to 2013 and has long been an ardent environmentalist. The computer climate model developed by his research team was one of two models on which the Charney Report was based, and was the main building block for many of the models used today. Although Hansen was careful in his early published papers to qualify the predictions of his model and to point out its limitations, in public he began to make highly exaggerated predictions of future warming, none of which have come true. This included testimony he gave at the 1986 Senate hearing.[2]

In 1988, the Senate Energy and Natural Resources Committee held another hearing on climate change, on a record hot day for Washington, DC of 37° Celsius (98° Fahrenheit). For many years it was widely believed that

---

[1] U.S. Senate Committee on Environment and Public Works. 1986. *Ozone Depletion, The Greenhouse Effect, and Climate Change*. Washington, DC: U.S. Government Printing Office, 1, http://njlaw.rutgers.edu/collections/gdoc/hearings/8/86602726a/86602726a_1.pdf. Accessed 12 July 2018.

[2] Ibid, 18-30, 78-97. Hansen's 1986 testimony to the subcommittee predicted average warming in the U.S. of about 1.5° Celsius (2.7° Fahrenheit) 30 years later by 2016, assuming that growth of the atmospheric CO_2 level would continue to accelerate at the same rate. The CO_2 growth rate subsequently increased slightly, a change that would have led to an even higher prediction. [U.S. Department of Commerce, National Oceanic and Atmospheric Administration. 2018. "Trends in Atmospheric Carbon Dioxide: Mauna Loa, Hawaii," http://www.esrl.noaa.gov/gmd/ccgg/trends/data.html. Accessed 12 July 2018.]

Yet the actual warming from 1986 to 2015, just before an El Niño temperature spike began, was only 0.5° Celsius (0.9° Fahrenheit) in the contiguous U.S., based on five-year LOWESS smoothed annual mean temperatures. [U.S. National Aeronautics and Space Administration, Goddard Institute for Space Studies. 2017. "GISS Surface Temperature Analysis: Annual Mean Temperature Change in the United States," http://data.giss.nasa.gov/gistemp/graphs/graph_data/U.S._Temperature/graph.txt. Accessed 12 July 2018.]

the committee chairman, Senator Timothy Wirth, had for dramatic effect scheduled the hearing on what was likely to be the hottest day of the summer, and had opened the windows of the room the night before, so that the air conditioning wasn't working during the hearing and witnesses needed to wipe their brows as they testified. Recently, however, former Senator Wirth admitted this story was a myth.[1] What's true, though, is Hansen's claim before the committee of 99% certainty from his computer simulations that the 0.4° Celsius (0.7° Fahrenheit) warming from 1958 to 1987 wasn't a natural variation but was caused primarily by the buildup of CO_2 and other man-made gases in the atmosphere. Even though many of his scientific colleagues didn't agree with him, Hansen's excesses nevertheless succeeded in raising public consciousness about global warming, both in the U.S. and elsewhere.

The year 1988 was also marked by a major international political event in global warming: the founding of the Intergovernmental Panel on Climate Change (IPCC), a powerful organization of climate scientists, government officials and bureaucrats that issues regular assessment reports on global warming and has become a leading advocate for the theory of man-made climate change. Establishment of the IPCC was the outcome of several earlier events triggered both by the ice age scare of the 1970s and by the burgeoning environmental movement.

In 1979, the same year that the Charney Report on CO_2 was published, the first World Climate Conference was held in Geneva. Convened by the World Meteorological Organization (WMO), a group that works to standardize weather observations, together with several United Nations agencies, the conference stated its fear that "continued expansion of man's activities on Earth may cause significant extended regional and even global changes of climate."[2] Although the Geneva meeting didn't call for any particular action, it led to two further conferences in Villach, Austria, in 1980 and 1985, specifically to examine the role of CO_2 and other greenhouse gases[3] in climate change. The 1985 Villach conference concluded that ever increasing greenhouse gases were likely to cause global warming in the first half of the 21st century "greater than in any man's history." The gathering, attended by

---

[1] Kessler, Glenn. 2015. "Setting the Record Straight: The Real Story of a Pivotal Climate-Change Hearing." *Washington Post*, 30 March, https://www.washingtonpost.com/news/fact-checker/wp/2015/03/30/setting-the-record-straight-the-real-story-of-a-pivotal-climate-change-hearing/. Accessed 12 July 2018.

[2] World Meteorological Organization. 1979. "Declaration of the World Climate Conference," http://www.dgvn.de/fileadmin/user_upload/DOKUMENTE/WCC-3/Declaration_WCC1.pdf. Accessed 12 July 2018.

[3] Greenhouse gases in Earth's atmosphere include water vapor (H_2O), carbon dioxide (CO_2), methane (CH_4), nitrous oxide (N_2O), ozone (O_3), chlorofluorocarbons (CFCs) and hydrofluorocarbons (HFCs).

delegates from 29 countries, was significant as it was the beginning of the close entanglement of climate science with politics, and it was the first time that such a meeting had emphasized urgency in dealing with possible future climate change.

Along with smaller meetings, the 1985 Villach conference was a major catalyst for the establishment of the IPCC in 1988. The stated purpose of the IPCC, founded jointly by the UN Environment Programme and the WMO, was to assess "the scientific, technical and socioeconomic information relevant for understanding the risk of human-induced climate change." To accomplish this aim, the organization divides its work between three working groups and an associated task force and task group. It's the job of Working Group I to review the science underlying global warming; Working Group II is more political, with a current mandate to assess the environmental and socioeconomic impacts of climate change; while Working Group III focuses on mitigation strategies.

Given its mandate, it was inevitable that the IPCC would be biased toward the assertion that humans have caused most of the warming measured since 1850. The IPCC's Second Assessment Report in 1995 was both controversial and ambiguous, declaring that the evidence "suggests a discernible human influence on global climate."[1] However, the Third Assessment Report in 2001, which featured the infamous "hockey stick" temperature chart,[2] confidently asserted the IPCC was 66% to 90% certain that most of the observed warming over the previous 50 years had come from increased greenhouse gas levels in the atmosphere. Quantitative estimates of the IPCC's confidence level were absent from the Fifth Assessment Report in 2013, but the report nevertheless proclaimed it "extremely likely that human

---

[1] IPCC. 1996. Summary for Policymakers. In J. T. Houghton, L. G. Meira Filho, B. A. Callander et al., eds., *Climate Change 1995: The Science of Climate Change. Contribution of Working Group I to the Second Assessment Report of the Intergovernmental Panel on Climate Change.* Cambridge: Cambridge University Press, 4.

[2] The hockey stick graph purported to show that the global temperature remained more or less the same over the past 1,000 years (the shaft of the stick), but took a sudden, rapid upturn (the blade) in the 20th century. The temperature reconstruction eliminated both the Medieval Warm Period and the Little Ice Age, which are well-documented features of our past climate. However, the hockey stick has since been debunked by a number of authors, including some who side with the IPCC on human-induced warming, as well as a team of scientists and statisticians assembled by the National Research Council of the U.S. National Academy of Sciences. [U.S. National Research Council, Board on Atmospheric Sciences and Climate. 2006. *Surface Temperature Reconstructions for the Last 2,000 Years.* Washington, DC: National Academies Press, chaps. 9 and 11, http://www.nap.edu/catalog/11676/surface-temperature-reconstructions-for-the-last-2000-years. Accessed 12 July 2018.]

influence has been the dominant cause of the observed warming since the mid-20th century."[1]

Science aside, formation of the IPCC accelerated the pace of political action on global warming, at a time when temperatures were still rising rapidly across the planet. In 1990, the same year that the IPCC issued its First Assessment Report, the second World Climate Conference called for a global treaty on climate change. This led to extensive negotiations among 150 countries prior to a so-called "Earth Summit" in Rio de Janeiro in 1992, at which the UN Framework Convention on Climate Change (UNFCCC) was signed; it has now been ratified by 197 parties.

The lofty goal of the treaty, which took effect in 1994, was to stabilize greenhouse gas concentrations in the atmosphere "at a level that would prevent dangerous anthropogenic [human-caused] interference with the climate system." Shrinking of greenhouse gas emissions was to be done only in ways that would not disrupt the global economy or food production, an objective that even the IPCC itself later admitted might not be attainable and referred to as a "double-edged sword."[2] Although the UNFCCC didn't define "dangerous," the IPCC's Second Assessment Report in 1995 wisely stated that interpretation of the term involved political judgement.[3] The UNFCCC also required industrialized nations to help developing nations meet their treaty obligations by providing financial support and technology, a stipulation that many in both developed and developing countries see as anti-capitalist and merely a wealth transfer mechanism.

The voluntary commitments of the UNFCCC signatories to reduce their greenhouse gas emissions were replaced by mandatory emission limits under the Kyoto Protocol, adopted in Kyoto, Japan in 1997, although it didn't come into effect until 2005. The Kyoto scheme established legally binding targets for industrialized countries, with the goal of lowering global greenhouse gas emissions by an average of 5% of 1990 levels, over the five years from 2008 to 2012. Industrialized nations were singled out because the current high concentrations of CO_2 and other greenhouse gases are largely the result of

---

[1] IPCC. 2013. Summary for Policymakers. In T. F. Stocker, D. Qin, G.-K. Plattner et al., eds., *Climate Change 2013: The Physical Science Basis. Contribution of Working Group I to the Fifth Assessment Report of the Intergovernmental Panel on Climate Change*. Cambridge: Cambridge University Press, 17.

[2] IPCC. 2007. Rogner, H-Holger, Dadi Zhou, Rick Bradley et al. Introduction. In B. Metz, O. R. Davidson, P. R. Bosch et al., eds., *Climate Change 2007: Mitigation. Contribution of Working Group III to the Fourth Assessment Report of the Intergovernmental Panel on Climate Change*. Cambridge: Cambridge University Press, 99.

[3] IPCC. 1996. Technical Summary. In R. T. Watson, M. C. Zinyowera, R. H. Moss and D. J. Dokken, eds., *Climate Change 1995: Impacts, Adaptations and Mitigation of Climate Change: Scientific-Technical Analyses. Contribution of Working Group II to the Second Assessment Report of the Intergovernmental Panel on Climate Change*. Cambridge: Cambridge University Press, 21.

industrial activity over the past 150 years. The U.S. signed, but never ratified, the protocol because China and India — who together emit over a third of the world's CO_2, and much more than the U.S. — were exempt from its requirements; Canada ratified the protocol but withdrew in 2012. Although the Kyoto Protocol was extended from 2012 to 2020, far fewer countries have so far ratified the extension than the original protocol.

Implementation of the overarching UNFCCC was put in the hands of an annual Conference of the Parties (COP). The Kyoto Protocol was crafted at COP-3 in 1997, while COP-21 in 2015 drew up what is known as the Paris Agreement. The avowed aim of this agreement was to strengthen the global response to the perceived threat of climate change by "holding the increase in the global average temperature to well below 2° Celsius (3.6° Fahrenheit) above preindustrial levels," preferably limiting the temperature increase to only 1.5° Celsius (2.7° Fahrenheit).[1] As we'll see later in the chapter, the belief that humans know enough about climate to regulate the earth's thermostat is preposterously unscientific, whether computer climate models exaggerate global warming or not.

In contrast to the legally binding limits on greenhouse gas emissions imposed on industrialized countries under the Kyoto Protocol, the Paris Agreement simply requires its 195 signatories to declare a voluntary "nationally determined contribution" toward emission reduction,[2] so has no real teeth. The agreement became effective in 2016, after meeting the specified condition of ratification by at least 55 parties to the UNFCCC that accounted for a minimum of 55% of total greenhouse gas emissions. The U.S. initially ratified the agreement this time, but in 2017 announced its intention to withdraw. And while China and India both ratified it, China declared its intention of *increasing* CO_2 emissions up until 2030. Nonetheless, it has been estimated that, even if we *could* regulate the climate and even if all countries follow through with their voluntary contributions, the actual mitigation of global warming by 2100 would be at most only 0.2° Celsius (0.4° Fahrenheit).[3]

But despite all this political effort to cut back on greenhouse gas production, the CO_2 level in the earth's atmosphere has continued to increase relentlessly. At the same time global warming has drastically slowed, almost

---

[1] United Nations. 2015. Paris Agreement, Article 2, http://unfccc.int/files/essential_background/convention/application/pdf/english_paris_agreement.pdf. Accessed 12 July 2018.

[2] Ibid, Article 4.

[3] Cass, Oren M. 2015. Testimony to U.S. Senate Committee on Environment and Public Works, Hearing on Examining the International Climate Negotiations, 4, https://www.epw.senate.gov/public/_cache/files/6658e59c-1098-4820-8360-3d61a34985bf/cass-testimony.pdf. Accessed 12 July 2018.

to a standstill, for nearly 20 years, which is one reason that skeptics dispute the notion that our CO_2 emissions have a major effect on the climate system. Most skeptics are not "deniers" — a derogatory label applied to them by believers in climate change orthodoxy — but rather "lukewarmers." Lukewarmers such as me accept that global warming is real, but not that it's entirely man-made or that it's dangerous. Lukewarmers therefore see no need for drastic action on CO_2 emissions.

Just like those who unreservedly accept the conclusions of the IPCC's assessment reports and attend COP meetings, skeptics also organize regular conferences. Skeptics' meetings question the impact of greenhouse gas emissions on our climate, and delve into natural sources of global warming such as solar variability and ocean cycles. Typical meetings are attended by hundreds of scientists, economists and policymakers, compared with the thousands who participate in the UN's COPs, although the vast majority of the latter attendees are bureaucrats and diplomats, not scientists. Because of the bias toward human-induced global warming inherent in the IPCC's assessment reports, skeptics have formed the Nongovernmental International Panel on Climate Change (NIPCC) that produces its own lengthy, nonpolitically motivated reports challenging the conclusions of IPCC reports.[1, 2]

One of the political weapons that warmists use to bolster their case for dangerous man-made warming is the supposed 97% consensus among climate scientists on the issue. The 97% number comes from two principal sources, the first being a 2008 online survey of 3,146 earth scientists, of whom just 77 identified themselves as actively publishing in climate research.[3] Out of the 77 climatologists, 75 (or 97%) answered yes to the question: "Do you think human activity is a significant contributing factor in changing mean global temperatures?" Apart from the tiny sample size, 77 being only a few percent of the thousands of climate scientists worldwide, the survey result is highly misleading because many lukewarmer skeptics, at least those who accept that up to 50% of observed warming comes from humans, would also have answered yes to the same question. So 97% is much too high an estimate.

[1] NIPCC. 2013. *Climate Change Reconsidered II: Physical Science*, C. D. Idso, R. M. Carter and S. F. Singer (eds.). Chicago: The Heartland Institute.

[2] NIPCC. 2014. *Climate Change Reconsidered II: Biological Impacts*, C. D. Idso, S. B. Idso, R. M. Carter and S. F. Singer (eds.). Chicago: The Heartland Institute.

[3] Doran, Peter T. and Maggie Kendall Zimmerman. 2009. "Examining the Scientific Consensus on Climate Change." *Eos, Transactions, American Geophysical Union*, 90, 22-23, http://onlinelibrary.wiley.com/doi/10.1029/2009EO030002/epdf. Accessed 12 July 2018.

The second source of the 97% consensus number is a far more extensive study of 11,944 abstracts of research papers on climate science published between 1991 and 2011.[1] Among the 4,014 abstracts that expressed a position on anthropogenic global warming, 97% were judged as endorsing the scientific consensus. The same percentage was found for 2,142 of the assessed papers, whose authors were contacted and invited to self-rate their papers. However, what is most telling is that 7,930 of the 11,944 abstracts took no position at all on anthropogenic warming, nor did 761 of the 2,142 papers by contacted authors. If *all* abstracts and papers are included in the calculations, the consensus percentage falls to between only 33% and 63%, respectively. These lower and more scientifically honest numbers are more in line with other estimates of the consensus among climate scientists.[2]

A measure of the consensus among scientists in general is a 2016 online survey of members of the American Meteorological Society (AMS); approximately one third of the participants considered themselves experts in climate science, though weren't necessarily climate scientists.[3] Of the 4,004 members who responded to a question on the cause of climate change, just 67% thought that recent warming was mostly or entirely due to human activity. The 33% who were skeptical and thought that natural variation plays a major role, or simply said they didn't know, isn't a majority but hardly the small minority that political activists on global warming insist that skeptics are. In the public at large, at least in the U.S., a 2018 Gallup poll pegged the percentage of global warming skeptics at 36%.[4] Other polls put the percentage of public skeptics even higher, especially in Europe.

Nevertheless, despite the scientific and public skepticism, a large number of the world's professional scientific societies and national academies of science have endorsed the IPCC stance on climate change, to a greater or lesser degree. Societies such as the American Chemical Society, the American Geophysical Union and the UK's Royal Society all declare

[1] Cook, John, Dana Nuccitelli, Sarah A. Green et al. 2013. "Quantifying the Consensus on Anthropogenic Global Warming in the Scientific Literature." *Environmental Research Letters*, 8, 024024, 1-7, http://iopscience.iop.org/article/10.1088/1748-9326/8/2/024024/pdf. Accessed 12 July 2018.
[2] Tol, Richard S. J. 2016. "Comment On 'Quantifying the Consensus on Anthropogenic Global Warming in the Scientific Literature' ." *Environmental Research Letters*, 11, 048001, 1-6, http://iopscience.iop.org/article/10.1088/1748-9326/11/4/048001/pdf. Accessed 12 July 2018.
[3] Maibach, Edward, David Perkins, Zephi Francis et al. 2016. "A 2016 National Survey of American Meteorological Society Member Views on Climate Change: Initial Findings," March. Fairfax, VA: George Mason University, https://gmuchss.az1.qualtrics.com/CP/File.php?F=F_cRR9lW0HjZaiVV3. Accessed 12 July 2018.
[4] Brenan, Megan and Lydia Saad. 2018. "Global Warming Concern Steady Despite Some Partisan Shifts," 28 March, https://news.gallup.com/poll/231530/global-warming-concern-steady-despite-partisan-shifts.aspx. Accessed 12 July 2018.

outright that warming over the last 50 years has come mostly from human activity, the same opinion expressed in the IPCC's Fifth Assessment Report. However, this view is not necessarily that of 100% of each society's members, who weren't polled although some were allowed to comment on drafts of the official statements. And even the position statement of the AMS, whose members were polled as discussed above, ignores the skeptical 33% in stating that: "Avoiding this future warming will require a large and rapid reduction in global greenhouse gas emissions."[1] In fact, many members of scientific societies have resigned their memberships over the issue of consultation on climate policy, saying that they felt steam-rolled by the society hierarchy.

Some organizations have taken more extreme positions. For example, the European Academies' Science Advisory Council, an alliance of the national science academies of European Union member states, finds that the IPCC's predictions about climate change are "overly conservative."[2] Yet the American Association of Petroleum Geologists, while admitting that its membership is divided on the influence of CO_2 emissions on global warming, says that future warming could well be less than documented natural variations in the past climate, and calls for expanded climate research. On the other hand, the principal concern among several Asian and African national academies is potential water shortages rather than CO_2 emissions.

But could so many learned societies and academies be wrong about global warming? The answer is an unqualified yes, because the world has harbored many such widespread scientific misconceptions before. We've already encountered two examples in this book. Chapter 1 discussed the famous example of Galileo, who fought the long-standing but mistaken consensus that the earth was the center of the solar system, and spent his last days sentenced to house arrest by the Inquisition for advocating the rival but later accepted sun-centered theory of Copernicus. The consensus on the earth-centered geocentric theory was upheld not only by the Catholic church but also by Aristotelian academics in universities of the time. And in Chapter 2, we saw how the stubbornly misguided consensus of the geology establishment about rigidly fixed continents delayed acceptance of Wegener's radical continental drift theory for at least 40 years.

Another example, from the field of medicine, is the discovery of the now unquestioned importance of handwashing to prevent infection in hospitals.

[1] American Meteorological Society. 2012. "Climate Change," https://www.ametsoc.org/ams/index.cfm/about-ams/ams-statements/statements-of-the-ams-in-force/climate-change/. Accessed 12 July 2018.

[2] European Academies' Science Advisory Council. 2015. "Facing Critical Decisions on Climate Change in 2015," http://www.leopoldina.org/uploads/tx_leopublication/2015_Easac_COP21_web.pdf. Accessed 12 July 2018.

In mid-19th-century European hospitals, up to 10% of women who had their babies delivered by obstetricians used to die from so-called childbed fever shortly after childbirth. In deliveries performed by midwives or at home, the death rate was several times lower, though no one understood why. But in 1847, a combative Hungarian obstetrician named Ignaz Semmelweis (1818–65) came up with the brilliant realization that hospital obstetricians weren't washing their hands after carrying out barehanded autopsies on the pus-ridden bodies of women who had suffered an anguished death from the same disease only the day before. Long before disease was associated with germs, Semmelweis deduced that the doctors were transferring some kind of "morbid poison" from the corpses to the women in labor.[1]

Once Semmelweis proposed that doctors wash their hands vigorously in a chlorinated lime solution after conducting autopsies, the mortality rate in obstetric wards plummeted. However, the discovery was widely rejected by the medical establishment of the day, whose consensus about disease was that it came from an imbalance in the four bodily humors of Hippocratic medical theory. It wasn't until 20 years later, after Semmelweis had been beaten to death in an asylum, that the newly accepted germ theory of disease vindicated his infection theory of childbed fever.

The medieval church, 19th-century medical authorities and the 20th-century geology establishment were all proved wrong and, as we saw in Chapter 4, the former consensus on dietary fat and heart disease is evaporating. So the conventional political wisdom on climate change could well be mistaken too. The earth is undoubtedly warming, but there's very little empirical evidence that humans are playing a major role.

Evidence That Doesn't Match Predictions

A crucial step in the scientific method is the comparison of observations with the predictions of a hypothesis, which itself is usually predicated on previous observations. In the case of global warming, the hypothesis is that the observed warming — which was approximately 0.85° Celsius (1.5° Fahrenheit) measured from 1880 to 2012 — has been caused primarily by ever-increasing human emissions of CO_2 and other greenhouse gases into the atmosphere. Computer climate models founded on the CO_2 hypothesis have been used to predict future hothouse conditions ever since the Charney Report in 1979, and are the basis of all the political statements made about the effects of greenhouse gases, such as the Paris Agreement. Current models

---

[1] Markel, Howard. 2015. "In 1850, Ignaz Semmelweis Saved Lives with Three Words: Wash Your Hands." *PBS NewsHour*, 15 May, http://www.pbs.org/newshour/updates/ignaz-semmelweis-doctor-prescribed-hand-washing/. Accessed 12 July 2018.

predict a future temperature from 1.5° Celsius (2.7° Fahrenheit) to 4.5° Celsius (8.1° Fahrenheit) higher than in 1850, once the CO_2 concentration reaches twice its preindustrial level[1] later this century, which is a similar but wider range than predicted in the Charney Report.

Not only do the climate models predict temperatures 100 years from now, but they also make projections of short-term warming that can be readily compared to actual observations recorded since the models were first developed in the 1970s. However, the short-term projections have failed badly, overestimating the warming rate by two or three times — a shortcoming admitted even by the IPCC. In the IPCC's Fifth Assessment Report, global warming in 2012 projected by 42 different climate models ranged from approximately 0.8° Celsius (1.4° Fahrenheit) to 1.5° Celsius (2.7° Fahrenheit) above the mean temperature for the period from 1850 to 1900. The actual observed warming in 2012, derived from global surface temperatures, was very close to the 0.8° Celsius (1.4° Fahrenheit) at the bottom end of this range, or 0.35° Celsius (0.65° Fahrenheit) below the mean of the model projected temperatures[2] That's a big discrepancy, and one that continued beyond 2012.

How well a particular model of any kind predicts reality is measured by what statisticians call predictive or forecast skill. Clearly, the predictive skill of computer climate models is poor, despite the misguided belief of IPCC report authors to the contrary.[3] Climate scientists who employ computer models to make forecasts of future climate conditions justify their confidence in the models through a process known as "hindcasting."

Hindcasting involves the use of a climate model to predict a known climatic feature in the recent past, such as the annual mean temperature or rainfall, with only observations prior to that time being fed into the computation. If the hindcast successfully reproduces the known temperature or rainfall distribution, climatologists consider the model skillful enough to

[1] The preindustrial level of CO_2 in the Earth's atmosphere was about 280 parts per million. [IPCC, 2013, 11.]

[2] Kirtman, Ben, Scott B. Power, Akintayo John Adedoyin et al. 2013. Near-Term Climate Change: Projections and Predictability. In T. F. Stocker, D. Qin, G.-K. Plattner et al., eds., *Climate Change 2013: The Physical Science Basis. Contribution of Working Group I to the Fifth Assessment Report of the Intergovernmental Panel on Climate Change.* Cambridge: Cambridge University Press, figs. 11.25 (a) and (b).

[3] Ibid, chapter 11 executive summary. Note, however, that the authors of Chapter 9 of the same report are less hubristic about the ability of climate models to predict the future. [Flato, Gregory, Jochem Marotzke, Babatunde Abiodun et al. 2013. Evaluation of Climate Models. In T. F. Stocker, D. Qin, G.-K. Plattner et al., eds., *Climate Change 2013: The Physical Science Basis. Contribution of Working Group I to the Fifth Assessment Report of the Intergovernmental Panel on Climate Change.* Cambridge: Cambridge University Press, chapter 9 executive summary.]

make reliable projections of the future climate. However, this conclusion is scientifically unwarranted because it assumes no change with time of internal climate processes poorly represented in the models, such as ocean cycles, or of external processes omitted from the models, such as certain types of solar variability. If any of these processes behave differently during some future period than during the period of the hindcast, the future projection won't be valid. In any case, computer climate models certainly fail the test of short-term projection, casting serious doubt on their longer term predictions as well — a highly important issue, given the enormous amounts of money and time that have been invested in the political efforts to put a lid on our greenhouse gas emissions.

The divergence of computer model predictions from measured temperatures only began in the late 1990s, when the warming surge that began in the 1970s came to an abrupt halt. After the global temperature soared by about 0.5° Celsius (0.9° Fahrenheit) over this period, an increase comparable to that between 1910 and 1940, the rise suddenly slowed to a crawl. But climate models had called for the mercury to keep going up at the same clip, as more and more CO_2 was released into the atmosphere: the predicted warming rate from 1998 to 2012 was a rapid 0.21° Celsius (0.38° Fahrenheit) every decade. The actual rate for this 14-year period, based on surface temperatures, was only a meager 0.04° Celsius (0.07° Fahrenheit) per decade. This is five times lower than predicted[1] and almost three times lower than the longer-term warming rate of 0.11° Celsius (0.20° Fahrenheit) per decade, or 1.1° Celsius (2.0° Fahrenheit) per century, from 1951 to 2012. Likewise, a comparison of observed warming derived from satellite temperature measurements, which only debuted in 1979, with warming predicted by 32 climate models reveals that the models typically warm the global atmosphere twice as fast as they should; only the prediction of a single Russian model comes close to the observations.[2]

The failure of real-world temperatures to match climate model predictions has become known as the global warming "pause" or "hiatus." Because of the 2015–16 El Niño, a natural Pacific Ocean cycle that causes a large temporary warming spike, we won't know until 2018 or 2019 whether the pause ended in 2015 or has continued. But despite acknowledgment

---

[1] The measured warming rate of 0.04° Celsius (0.07° Fahrenheit) per decade from 1998 to 2012 may be artificially low because of the selected starting year: global temperatures temporarily spiked in 1998, due to the unusually strong El Niño of 1997-98.

[2] Christy, John R. 2016. Testimony to U.S. House of Representatives Committee on Science, Space & Technology, Hearing on Paris Climate Promise: A Bad Deal for America, 12-13, https://science.house.gov/sites/republicans.science.house.gov/files/documents/ HHRG-114-SY-WState-JChristy-20160202.pdf. Accessed 12 July 2018.

of the slowdown in the IPCC's 2013 report, as just noted, a small group of advocates for the politically motivated theory of catastrophic man-made warming insisted in several 2015 research papers that the pause had never occurred. This refusal to accept sound scientific evidence induced several well-known climate scientists to publicly admonish their colleagues in a 2016 paper; the chastising authors even included Michael Mann of hockey stick infamy.[1] The controversy over existence of the pause continued into 2017, with uncertainty as to whether warming from the recent El Niño had dissipated yet.

The pause is a major stumbling block for the CO_2 hypothesis of global warming. When actual empirical observations don't confirm the predictions of a hypothesis, the scientific method demands that the hypothesis be either thrown out, or modified to fit the evidence. Dozens of explanations for the pause, some farfetched, have been proposed over the last few years. Among these are three possible reasons set out in the IPCC's Fifth Assessment Report in 2013, in which the authors of the chapter on climate models are scientifically honest enough to admit that the CO_2-based models indeed need modification.

Of the reasons the authors advance to explain the hiatus in warming since 1998, the first is a cooling effect from some type of natural variability not properly incorporated in the models. Sources of natural variability include ocean cycles such as El Niño (only some elements of which are correctly simulated in the models) and the Pacific Decadal Oscillation, as well as solar variations other than the regular 11-year cycle.[2] The second reason put forward in the report is a combination of minor volcanic eruptions, the ash from which causes global cooling, together with the dimming phase of the solar cycle. The third IPCC explanation for the hiatus is overestimation by the models of the climatic response to CO_2 and other greenhouse gases, which we'll look at shortly, although the IPCC researchers say that this possible model error can't fully account for the observed slowdown in warming. Whether or not any or all of these possibilities, which were endorsed in a detailed analysis published by other authors in 2017,[3] can adequately explain the pause will depend on how long it lasts.

[1] Fyfe, John C., Gerald A. Meehl, Matthew H. England et al. 2016. "Making Sense of the Early-2000s Warming Slowdown." *Nature Climate Change*, 6, 224-228.
[2] The length of the solar cycle fluctuates from 9 to 14 years, though is normally about 11 years. The sun brightens very slightly during the first half of the cycle and then dims during the second half.
[3] Santer, Benjamin D., John C. Fyfe, Giuliana Pallotta et al. 2017. "Causes of Differences in Model and Satellite Tropospheric Warming Rates." *Nature Geoscience*, advance online publication, 19 June, doi:10.1038/ngeo2973, 1-11.

The poor predictive skill of climate models certainly indicates that something is missing from the models. Climate models are highly sophisticated computer programs that simulate a host of complex interactions in the earth's climate system, which embraces the atmosphere, landmasses, oceans, snow and ice, and the global ecosystem. Some of the diverse phenomena modeled in current computer programs include jet streams in the upper atmosphere, greenhouse gases, clouds, volcanic and man-made aerosols, deep ocean currents, sea ice and the earth's carbon cycle.

For serving up a picture of the likely climate a century from now, an important feature of a computer climate model is what are known as feedbacks; I've discussed climate models and feedback in another book.[1] It's feedbacks that govern climate sensitivity — defined as the amount of global warming that results from a doubling of CO_2 from its 1850 level, which as we've seen is projected to be from 1.5° Celsius (2.7° Fahrenheit) to 4.5° Celsius (8.1° Fahrenheit). The higher the climate sensitivity, the more global warming occurs, while the lower the sensitivity, the less warming computer models predict. Feedback, a technical term borrowed from electronic engineering, can be either positive or negative. Positive feedbacks amplify global warming, but negative feedbacks tamp it down.

What most people don't know is that without feedbacks, the climate would be so insensitive to CO_2 that global warming wouldn't be a concern. A doubling of CO_2, acting entirely on its own, would only raise global temperatures by 1.1° Celsius (2.0° Fahrenheit). According to climate modelers, atmospheric CO_2 needs assistance from another greenhouse gas, water vapor, in order to become troublesome at all. In climate models, it's positive feedback from water vapor — by far the most abundant greenhouse gas — and, to a lesser extent, feedback from clouds, snow and ice, that boosts the warming effect of doubled CO_2 alone from 1.1° Celsius (2.0° Fahrenheit) to the IPCC's elevated 1.5° Celsius (2.7° Fahrenheit) to 4.5° Celsius (8.1° Fahrenheit) range.

The assumption that water vapor feedback is positive and not negative was originally made by Arrhenius over a century ago, and forms a central part of the Charney Report. On the face of it, the assumption seems reasonable. The feedback arises when slight CO_2-induced warming of the earth causes more water to evaporate from oceans and lakes, and the extra moisture then adds to the heat-trapping water vapor already in the atmosphere. This amplifies the warming even more. But the magnitude of the feedback is

---

[1] Alexander, Ralph B. 2012. *Global Warming False Alarm: The Bad Science Behind the United Nations' Assertion that Man-Made CO_2 Causes Global Warming*. Royal Oak, MI: Canterbury Publishing, chaps. 3 and 4.

critically dependent on how much of the extra water vapor ends up in the upper troposphere[1] as the planet heats up, because it's the upper troposphere from where heat escapes to outer space; an increase in moisture there means stronger water vapor feedback and more heat trapping.

Unfortunately for climate science, the observational evidence on upper tropospheric humidity is sparse. The limited data available do show that upper troposphere humidity strengthened slightly in the tropics during the 30-year period from 1979 to 2009, which includes most of the $0.5°$ Celsius ($0.9°$ Fahrenheit) warming from the 1970s until 1998 and the first 10-15 years of the pause. However, it also *diminished* in the subtropics and possibly at higher latitudes also during this time.[2] The less water vapor in the upper troposphere, the less positive the water vapor feedback and the lower the climate sensitivity, which measures the climate's response to rising CO_2. As mentioned earlier, overly high sensitivity to CO_2 may be one reason that computer climate models were unable to predict the global warming pause.

It's much the same story for clouds, although the simulation of clouds is a weak point in climate models and even acknowledged by the IPCC as "challenging." Most computer climate models make assumptions about clouds that lead to positive cloud feedback, although a small number of models predict negative feedback. Just as with humidity, there's a paucity of meaningful observational data for clouds.

But if cloud feedback were actually negative instead of positive, and the water vapor feedback weren't as strongly positive as climate models indicate, the net climate feedback could be either close to zero or even negative, and man-made global warming would be insignificant. The whole political narrative about greenhouse gases and dangerous anthropogenic global warming depends on computer models based on *theoretical* assumptions about feedbacks and climate sensitivity. Science, remember, takes its cue from *observational* evidence.

Apart from feedbacks, another major deficiency in climate models is their inability to simulate natural patterns in the climate system. As one example, the models fall down in predicting the timing and climatic effects of several

---

[1] The troposphere is the lowest layer of the earth's atmosphere, and the layer where most of our weather occurs.

[2] Shi, Lei and John J. Bates. 2011. "Three Decades of Intersatellite-Calibrated High-Resolution Infrared Radiation Sounder Upper Tropospheric Water Vapor." *Journal of Geophysical Research*, 116, D04108, 1-13.

The authors found that upper troposphere specific humidity increased by 0.25% per decade from 1979 to 2009 in the tropics ($10°$ in latitude either side of the equator), while it decreased by 0.2% per decade over the same period in the subtropics (from $20°$ to $30°$ either side of the equator). IPCC data for humidity throughout the troposphere suggests that upper troposphere humidity also decreased at higher latitudes.

---

major ocean cycles such as the Atlantic Multidecadal Oscillation, the Pacific Decadal Oscillation and the closely related Interdecadal Pacific Oscillation. All these and other ocean cycles repeat at intervals ranging from years to decades and are thought to play a big role in regulating the planetary climate. But, even though oceans cover 71% of the earth's surface and hold most of the heat from global warming, the IPCC has only low to medium confidence in the capacity of computer climate models to reproduce the dominant features of two thirds of the known ocean cycles.[1] Again, the models depend more on theoretical assumptions than on actual scientific evidence.

An encouraging step to correct this imbalance, however, is a new program initiated by the U.S. National Oceanic & Atmospheric Administration (NOAA) and NASA to study the El Niño cycle as it unfolds. In early 2016, at the peak of one of the strongest El Niños on record, a research aircraft with a team of scientists on board flew close to the heart of the disruptive event.[2] Their measurements were complemented by remote observations from a robotic aircraft and weather balloons, in the hope of obtaining better data to feed into climate and weather-forecasting models.

A further natural source of variability is possible indirect solar effects. Present climate models mostly include just the direct effect of the sun's radiation on the earth's climate, which causes only a few percent of total global warming according to the models. While this estimate may be on the low side, indirect solar effects due to the sun's ultraviolet (UV) radiation or cosmic rays from deep space could also contribute to global warming, although some sort of amplification mechanism would be needed, analogous to the assumed amplification by water vapor and cloud feedbacks that boosts the modest temperature increase caused by CO_2 acting solo.[3] Two positive feedback mechanisms that could explain local climatic trends in the Indo-Pacific region, and are thought to result from indirect solar effects, have been successfully simulated in some climate models. But empirical evidence for any global trends resulting from indirect solar effects is limited so far, which creates little incentive to incorporate them in the models.

Nonetheless, it has been predicted that direct solar effects not in play at present may cause global cooling in the near future, a drastic potential change that is definitely not foreseen by current computer climate models. The prediction follows from the observation of sunspots, which are small dark blotches on the sun caused by magnetic storms on the sun's surface.

---

[1] Flato, Marotzke, Abiodun et al., 2013, table 9.4.
[2] U.S. Department of Commerce, National Oceanic & Atmospheric Administration. 2016. "Unprecedented Effort Launched to Discover How El Niño Affects Weather," 5 February, http://www.esrl.noaa.gov/psd/news/2016/020516.html. Accessed 12 July 2018.
[3] Alexander, 2012, 84-89.

Along with the sun's heat and light, the average monthly number of sunspots waxes and wanes during the solar cycle. Recently, the peak number of sunspots seen in a cycle has been declining, prompting expectations that the number will soon drop to nearly zero — a phenomenon that last occurred during the so-called Maunder Minimum, a 70-year cool period in the 17th and 18th centuries that was part of the Little Ice Age. Two hypotheses, one based on a 210-year solar cycle,[1] the other on an anticipated drop in the sun's magnetic field,[2] project a repeat Maunder Minimum beginning around 2020, but perhaps only 35 years long. Whether the subsequent cooling from such an unexpected solar event will dominate the ongoing warming from feedback-assisted CO_2, as seems quite likely, or whether CO_2 warming will gain the upper hand as the 2013 IPCC report anticipates, is merely speculation for now. Science requires us to wait for the evidence before drawing any conclusions.

As discussed earlier, the soundness or predictive skill of climate models is evaluated through hindcasting. Hindcasting is also used to simulate the "fingerprint patterns" of various possible causes of global warming: if a simulated fingerprint pattern, or combination of patterns, matches observed climate changes, climatologists attribute the changes to the causes characterized by those patterns. This procedure is behind the assertion in the 2013 IPCC report that human activities have caused more than half the global warming from 1950 to 2010. Specifically, the report claims that the escalation in temperature over this interval could only have been caused by a combination of human-induced changes in greenhouse gases and aerosols, together with natural variability; simulations driven only by natural causes don't reproduce observed temperatures, the report says. But this IPCC claim about fingerprints is completely undercut by the existence of the global warming slowdown from 1998 onwards. While the fingerprint pattern for global temperatures appears to be a good match to the observational evidence from 1970 until 1998, if fails dismally after that, and is a poor representation of earlier warming and cooling periods.

That computer climate models have so many weaknesses should hardly be surprising. The earth's climate is extremely complex, and climate science is still in its infancy. Despite having put a man on the moon, and having developed thermonuclear weapons powerful enough to destroy

[1] Abdussamatov, Habibullo I. 2012. "Bicentennial Decrease of the Total Solar Irradiance Leads to Unbalanced Thermal Budget of the Earth and the Little Ice Age." *Applied Physics Research*, 4, 178-184.
[2] Shepherd, Simon J., Sergei I. Zharkov and Valentina V. Zharkova. 2014. "Prediction of Solar Activity from Solar Background Magnetic Field Variations in Cycles 21-23." *Astrophysical Journal*, 795: 46, 1-8.

humanity, we still have only a rudimentary scientific understanding of climate. There are many gaps in the theory that underpins climate models, and climate scientists must therefore make a lot of guesses to fill in the gaps. To pretend that present models truly represent the real world is sheer arrogance on our part.

The Multiple Lines of Evidence Fallacy

A scientific hypothesis does not become a full-fledged theory until it's been tested multiple times and perhaps modified. In most cases, a single line of evidence isn't enough to establish a scientific consensus about the theory; widespread consensus requires multiple independent lines of evidence, as already discussed in this book. Acceptance of Darwin's theory of evolution, for example, awaited the convergence of supporting evidence from a number of disparate fields such as paleontology (the fossil record), geology, biogeography, genetics, molecular biology and others. For dietary fat, on the other hand, conflicting lines of evidence from the fields of epidemiology, preventive medicine, nutrition science and statistics have had exactly the opposite effect, and the diet–heart hypothesis is being overturned.

In climate science, proponents of the hypothesis that humans are changing the earth's climate claim the hypothesis is supported by multiple lines of evidence. But even though it's true that several types of observation — of global temperature, sea levels, ocean heat, the atmosphere and ice melting — all demonstrate that the planet is heating up, the observations alone don't show that global warming is caused by humans. The only way that global warming can be tied to human activity is through computer climate models that depend on CO_2 to produce the warming. Yet these are the same models, as discussed in the previous section, that failed to predict the global warming pause, may overestimate water vapor feedback and thus climate sensitivity, and don't have a good handle on internal climate variability.

Apart from this tenuous connection between supposed cause and effect, the scientific evidence itself is often weak or questionable, as I'll discuss below. In a court of law, the evidence would certainly not be strong enough to prove beyond a reasonable doubt that climate change is anthropogenic, if the judge didn't toss out the case to begin with. Indeed, a British judge in 2007 ruled that "An Inconvenient Truth," an environmental documentary on climate change made by former U.S. Vice President Al Gore, and slated for showing in British high schools, contained no fewer than nine scientific errors.[1] In 2013, a challenge was mounted in the U.S. Supreme Court to Environmental Protection Agency (EPA) regulations that rely on the supposed multiple lines of evidence for climate change caused by human greenhouse gas emissions.[2] Although the Supreme Court upheld the EPA's authority to regulate emissions, it sidestepped the evidence question and ruled instead that the agency had overstepped its bounds in applying the regulations to small emission sources such as retail stores and schools.

Surface Temperature

Of all the evidence for global warming, the most well established and complete is the global temperature record since 1880, which is shortly after standardized methods for measuring temperatures were adopted. The record clearly shows the two periods, from 1910 to 1940 and from the early 1970s to 1998, when the mercury shot up rapidly, as well as the cooling spells from 1880 to 1910 and 1940 to 1970, together with the pause since 1998. Global surface warming over the whole interval from 1880 to the present is reflected in both land and sea observations, as well as satellite temperature data. However, even though it's quite certain that warming has occurred, its exact magnitude can justifiably be questioned, for two reasons.

The first reason is what we call the urban heat island effect. The effect describes land temperatures that show false warming because thermometers are soaking up heat from artificial sources such as a nearby air conditioning fan or asphalt paving, instead of sampling the true air temperature. The most recent IPCC report estimates that the heat island effect raises measured land temperatures by 10% to 15%.[3] But because landmasses cover only 29% of the earth's surface, the contribution of the urban heat island effect to global temperatures is less than 5%. Nonetheless, this small difference reveals itself in global satellite temperature measurements, which sample both land and sea and show a global warming rate slightly lower than that derived from surface thermometers.

[1] BBC News. 2007. "Gore Climate Film's Nine 'Errors' ," 11 October, http://news.bbc.co.uk/2/hi/uk_news/education/7037671.stm. Accessed 12 July 2018.

[2] U.S. Supreme Court, Brief of Amici Curiae Scientists and Economists in Support of Petitioner Southeastern Legal Foundation, Inc., et al. and State Petitioners. 2013. Utility Air Regulatory Group v. Environmental Protection Agency, et al., 573 U.S. 1, https://wattsupwiththat.files.wordpress.com/2013/12/amicus_curiae-ef_sc_merit_12-1146etseqtsacscientistsfinal_final.pdf. Accessed 12 July 2018.

[3] Hartmann, Dennis L., Albert M. G. Klein Tank, Matilde Rusticucci et al. 2013. Observations: Atmosphere and Surface. In T. F. Stocker, D. Qin, G.-K. Plattner et al., eds., *Climate Change 2013: The Physical Science Basis. Contribution of Working Group I to the Fifth Assessment Report of the Intergovernmental Panel on Climate Change.* Cambridge: Cambridge University Press, section 2.4.1.3.

A second reason that warming rates are probably inflated is adjustments to the raw data made by the custodians of global temperature measurements. The three principal guardians of the world's temperature data are NOAA and NASA's GISS in the U.S., and the UK collaboration (known as HadCRU) between the Climatic Research Unit at the University of East Anglia and the Met Office's Hadley Centre. Their practice of making adjustments to already recorded temperatures is an attempt to eliminate various biases that exist in the surface temperature record. These biases include the urban heat island effect, which causes land temperatures to be overstated as just mentioned; a switch in daily measuring times of maximum and minimum temperatures during the 1900s in the U.S., from the afternoon to morning when it's cooler, which results in understated recent temperatures; the upgrading of technology for land measurements from old-fashioned glass thermometers to more modern electronic sensors, which results in lower recorded maximums and higher recorded minimums; weather station moves, which can either raise or lower the recorded temperature;[1] and a gradual improvement in methods for measuring sea surface temperatures, with decreasing emphasis on ship observations and the introduction of measurements from moored and drifting buoys.

The result of all these adjustments is that the long-term warming rate on land has been boosted, while the rate for sea-based measurements has been lowered. But over the short term, for example from 1975 to 2015 or 1998 to 2015, the land surface warming rate is barely affected by the adjustments — with the notable exception of NOAA's data that show distinctly faster warming on land over these two periods than either GISS or HadCRU.[2] The adjusted sea warming rate computed by NOAA and GISS is also considerably higher than HadCRU for the period from 1998 to 2015.[3] In 2018, HadCRU's sea surface temperatures were lower than they had been five years earlier. The higher warming rates calculated by NOAA have led critics to accuse the agency of exaggerating global warming by excessively cooling the past and warming the present, a charge that may stem from their particular method

[1] Menne, Matthew J., Claude N. Williams Jr. and Russell S. Vose. 2009. "The U.S. Historical Climatology Network Monthly Temperature Data, Version 2." *Bulletin of the American Meteorological Society*, 90, 993-1007.

[2] Tisdale, Bob. 2016. "Updated: Do the Adjustments to Land Surface Temperature Data Increase the Reported Global Warming Rate?" *Watts Up With That* blog, 24 April, https://wattsupwiththat.com/2016/04/24/updated-do-the-adjustments-to-land-surface-temperature-data-increase-the-reported-global-warming-rate/. Accessed 12 July 2018.

[3] 2016. "Do the Adjustments to Sea Surface Temperature Data Lower the Global Warming Rate?" *Bob Tisdale — Climate Observations* blog, 9 April, https://bobtisdale.wordpress.com/2016/04/09/do-the-adjustments-to-sea-surface-temperature-data-lower-the-global-warming-rate/. Accessed 12 July 2018.

of adjustment but nevertheless raises questions about possibly politically motivated efforts to generate data in support of catastrophic anthropogenic warming. One of the 2015 research papers mentioned earlier that contended the global warming pause never existed, an assertion that was rebuked by other climate scientists, relied on NOAA's adjusted temperature data for its dramatic claim.

Any exaggeration of global warming by the urban heat island effect or possible data manipulation means that the IPCC's fingerprint pattern discussed in the preceding section isn't as close a fit to observed global temperatures from 1970 to 1998 as it seems. And, even if the exaggeration is ignored, the fingerprint pattern doesn't match the temperature record after 1998, as we've already seen. So while higher surface temperatures are indeed evidence of global warming, they're not necessarily a very strong line of evidence for a human link to climate change.

Atmospheric Temperature

A highly controversial line of evidence is the temperature variation in the earth's atmosphere, from the surface up through the troposphere to the stratosphere,[1] where it's cooler than on the ground. Computer climate models predict that global warming should heat the lower troposphere faster than the earth's surface, and the upper troposphere faster than the lower troposphere — the fingerprint of the latter being the so-called tropospheric "hot spot." Any source of global warming will produce a hot spot but, for CO_2 greenhouse warming, the warming rate of the air at an altitude of 10 to 12 kilometers (6 to 7 miles) directly above the tropics, where the difference is most conspicuous, should be about twice as large as it is near the surface, according to the models.[2] The predicted effect is much stronger for rising CO_2 levels than for other sources of global warming such as enhanced solar activity.

But, try as they might, atmospheric scientists have been unable to find the CO_2 hot spot. Although measurements of temperature aloft by either satellite or weather balloon are difficult and subject to large uncertainties,

---

[1] The stratosphere is the second lowest layer of the atmosphere, just above the troposphere.
[2] Bindoff, Nathaniel L., Peter A. Stott, Krishna Mirle AchutaRao et al. 2013. Detection and Attribution of Climate Change: From Global to Regional. In T. F. Stocker, D. Qin, G.-K. Plattner et al., eds., *Climate Change 2013: The Physical Science Basis. Contribution of Working Group I to the Fifth Assessment Report of the Intergovernmental Panel on Climate Change.* Cambridge: Cambridge University Press, section 10.3.1.2.1.
  Figure 10.8(b) shows an average model warming rate of approximately 0.4° Celsius (0.7° Fahrenheit) per decade from 1961 to 2010 at an altitude of 10-12 kilometers (6-7 miles), and an average rate of approximately 0.2° Celsius (0.4° Fahrenheit) per decade close to ground level.

the data indicate that the warming rate at the hotspot altitude is no different than it is closer to the ground, and both rates are less than climate models predict.[1] Satellite measurements are subject to error because the satellite temperature signal in the troposphere includes a small contribution from the stratosphere above, which needs to be subtracted; and weather balloon data are less precise than satellite temperature data. Nevertheless, hundreds of measurements have all confirmed that the warming rate in the upper troposphere is not just lower than predicted by climate models, but may even be lower than the surface warming rate.[2]

The models do a little better in the stratosphere where they predict cooling. Most of the cooling comes from depletion of ozone, a greenhouse gas, about 90% of which sits in the stratospheric ozone layer and about 10% in the troposphere. The ozone layer protects life on Earth by absorbing most of the sun's harmful UV radiation, which warms the lower stratosphere.[3] But in the late 1970s, scientists discovered that the layer was slowly being destroyed by human emissions of chlorofluorocarbons, the gases formerly used as refrigerants and in aerosol spray cans. This disturbing discovery led to the 1987 Montreal Protocol, an international treaty to phase out production of ozone depleting chemicals, which was also used as a template by environmental activists for subsequent agreements on climate change. Loss of ozone from the ozone layer due to human activity thins the layer, reducing the amount of solar radiation absorbed. The resulting cooling of the lower stratosphere has been documented since 1979, although the downswing has been interrupted by brief warming bursts from volcanic ash flung high into the stratosphere.[4] Computer climate models, however, underestimate the cooling, despite very accurate measurements of stratospheric ozone for the last 50 years.

If not the kiss of death, these findings certainly don't support the claim that atmospheric temperatures are evidence for human-induced climate change. The absence of a hot spot in the upper troposphere alone disqualifies

---

[1] Flato, Marotzke, Abiodun et al., 2013, section 9.4.1.4.2. Over the period from 1979 to 2012, the observed warming rate was between 0.02° Celsius (0.04° Fahrenheit) and 0.12° Celsius (0.22° Fahrenheit) per decade in the tropical middle and upper troposphere, and between 0.06° Celsius (0.11° Fahrenheit) and 0.13° Celsius (0.23° Fahrenheit) per decade in the tropical lower troposphere. Even the upper end of these two ranges falls short of the lower end of the respective ranges predicted by climate models.

[2] Hartmann, Klein Tank, Rusticucci et al., 2013, figure 2.27 and table 2.7.

[3] The ozone layer absorbs almost all the sun's UVB radiation, which causes sunburn and, at high doses, skin cancer. Absorption of solar UV warms the ozone layer and thus the stratosphere. But as a greenhouse gas, ozone slightly cools the stratosphere and warms the troposphere.

[4] The ash from volcanic eruptions results in cooling of the troposphere and the earth's surface, but warming of the stratosphere.

tropospheric temperature as a valid line of evidence for anthropogenic global warming. And the underestimated stratospheric cooling from ozone depletion is hardly a convincing argument for the soundness of the climate models needed to link global warming to human activity. The issue has been controversial, however, with a small band of climatologists insisting that a human influence on atmospheric temperatures can indeed be demonstrated, through a statistical analysis of the satellite data and model predictions for the troposphere.[1] But the evidence speaks for itself.

Ocean Heat

A third line of evidence cited for anthropogenic climate change is the heat stored in the world's oceans, known as the ocean heat content. The oceans play a key role in regulating global temperatures because they can hold a lot more heat — about 1,000 times more — than the atmosphere. Over 90% of the heat that global warming has added to the earth's climate system since 1970 resides in the oceans, manifesting itself as warming down to a depth of at least 700 meters (2,300 feet).[2] But, apart from sea surface temperatures, the temperature record for the oceans is sparse, making it difficult to compare climate model predictions with the evidence.

Not surprisingly, the strongest warming is found near the surface. In the uppermost 75 meters (250 feet), the average warming from 1971 to 2010 was 0.11° Celsius (0.20° Fahrenheit) per decade, comparable to the sea surface warming rate; 700 meters (2,300 feet) below the surface, it was an order of magnitude smaller at 0.015° Celsius (0.027° Fahrenheit) per decade. At greater ocean depths, it has only been possible to measure warming rates reliably since 2005, which is when a modern array of robotic diving buoys became extensive enough to cover all the world's oceans.

Five analyses have been made of the available ocean temperature data in order to estimate ocean heat content, which has steadily increased with time as we would expect in a warming world. Three of these five estimates show the same hiatus in ocean heat content as observed in global temperatures, except that the ocean heat slowdown is delayed by five years, starting in

---

[1] Santer, Benjamin D., Jeffrey F. Painter, Céline Bonfils et al. 2013. "Human and Natural Influences on the Changing Thermal Structure of the Atmosphere." *Proceedings of the National Academy of Sciences*, 110, 17235—17240.

[2] Rhein, Monika, Stephen R. Rintoul, Shigeru Aoki et al. 2013. Observations: Ocean. In T. F. Stocker, D. Qin, G.-K. Plattner et al., eds., *Climate Change 2013: The Physical Science Basis. Contribution of Working Group I to the Fifth Assessment Report of the Intergovernmental Panel on Climate Change*. Cambridge: Cambridge University Press, executive summary.

2003 rather than 1998.[1] Although the other two estimates don't reveal any slowdown, the hiatus is also clearly visible in sea surface temperatures after 1998. However, computer climate models fail to capture this hiatus in rising ocean heat, at least since 2005 when the more accurate diving buoy measurements began. For the upper ocean down to 700 meters (2,300 feet), the mean heat content projected by 15 different climate models exceeds the observation-based estimates, while for depths between 700 meters (2,300 feet) and 2,000 meters (6,600 feet) the projected heat content is less than the observations.[2] Just like global temperature data after the pause commenced, ocean heat content doesn't show the fingerprint of human-induced warming either.

Sea Levels

By far the most publicized phenomenon taken as evidence for a human influence on climate is rising sea levels, with the media regularly trumpeting some new prediction of the oceans flooding or submerging coastal cities in the decades to come. No one questions the fact that the average global sea level has been getting higher ever since the world started to warm after the Little Ice Age ended around 1850. Because water expands and takes up more volume as it gets warmer, higher ocean temperatures raise sea levels; levels also increase as glaciers and ice caps melt. But, as with surface temperatures, the precise magnitude of sea level rise and projections for the future have been questioned, especially since 1993, when satellite altimetry took over from tide gauges as the principal method of measurement.

According to the IPCC, the global rate of sea level rise from 1900 to 2010 was about 1.7 mm (about 1/16th of an inch) per year, accelerating after 1993 to almost double that rate.[3] However, there's some evidence that the rate of rise slowed down again after 2003, a slowdown that could be associated with the pause in rising ocean heat just discussed. But the satellite and tide gauge measurements are in stark disagreement and, in any case, the global rate is very much an average. Actual rates of sea level rise vary widely across the planet, and range from negative in Stockholm, corresponding to a falling sea level (as the region continues to rebound after melting of the last ice age's heavy ice sheet), to positive rates three times higher than the average in the

---

[1] Ibid, sections 3.2.3 and 3.2.4. Mathematically, the ocean heat content in a particular depth interval is the integral of the measured ocean temperature between the lower and upper depth limits, multiplied by the product of seawater density, seawater specific heat and ocean area.

[2] Gleckler, Peter J., Paul J. Durack, Ronald J. Stouffer et al. 2016. "Industrial-Era Global Ocean Heat Uptake Doubles in Recent Decades." *Nature Climate Change*, 6, 394-398.

[3] Rhein, Rintoul, Aoki et al., 2013, section 3.7.2.

western Pacific Ocean. And not only does sea level rise vary regionally, but it also fluctuates considerably over time. The rate was much higher than the 20th century average from 1920 to 1950, and from 1993 to 2003 as just noted, but much lower from 1910 to 1920 and 1955 to 1980[4] — periods that correspond roughly, with a 10-year to 20-year time delay, to last century's global warming and cooling stretches, respectively.

In an effort to connect sea level rise with human emissions of CO_2, computer model simulations have been made of both sea level and the rate of rise since 1900. The models include the thermal expansion of seawater as well as glacier melt, but no contributions from melting of the Greenland and Antarctic ice sheets, which are difficult to estimate accurately. However, the mean of 13 different computer models only comes within 20% of matching sea level observations from 1900 to 2010, and doesn't accurately reproduce the temporal fluctuations in the rate of sea level rise.[5] Better agreement with sea level observations is obtained by adding the observations of ice sheet melting (not included in the models) to the model predictions, but matching predictions plus ice sheet observations to measurements of sea level alone is hardly a valid comparison. Sea levels are, therefore, only a weak line of evidence for a human connection to climate change.

Sea and Land Ice

A further observation purported to be evidence for human-induced global warming is melting of ice, on both sea and land. The earth's two biggest ice sheets cover the polar landmasses of Antarctica and Greenland; sea ice, which expands during the winter and shrinks during the summer months, is found both in the Arctic and around Antarctica. Satellite measurements of Arctic and Antarctic sea ice have been made since 1979, while ice sheet data for Antarctica and Greenland have been gathered only since 1992.

What attracts the most attention is the loss of sea ice in the Arctic, where global warming has led to a dramatic reduction in ice cover. During September when the ice is at its minimum extent, the decrease in 2017 from the 1981 to 2010 median was approximately 27%.[6] However, this reduction was no less than it was in 2008, with the mean September coverage averaging

---

[4] Rhein, Rintoul, Aoki et al., 2013, section 3.7.4.
[5] Church, John A., Peter U. Clark, Anny Cazenave et al. 2013. Sea Level Change. In T. F. Stocker, D. Qin, G.-K. Plattner et al., eds., *Climate Change 2013: The Physical Science Basis. Contribution of Working Group I to the Fifth Assessment Report of the Intergovernmental Panel on Climate Change.* Cambridge: Cambridge University Press, section 13.3.6.
[6] U.S. National Snow and Ice Data Center. 2018. "Charctic Interactive Sea Ice Graph," http://nsidc.org/arcticseaicenews/charctic-interactive-sea-ice-graph/. Accessed 12 July 2018.

a 25% decrease over the 10-year period in between, suggesting that Arctic summer ice extent may also have paused, just like surface temperatures and ocean heat. And the very presence of summer ice at the North Pole refutes numerous prognostications by advocates of catastrophic man-made warming that the ice would be gone completely by 2016. Another fallacious claim about disappearing sea ice in the Arctic, one that has captured the public imagination like no other, is that the polar bear population would diminish along with the ice. But, while this may yet happen in the future, current evidence shows that the bear population has been stable for the whole time that the amount of ice has been measured.[1]

The Antarctic is a different story. Despite the contraction of sea ice in the Arctic, the sea ice around Antarctica has been expanding for more than 30 years, gaining in extent by an average of 1.8% per decade, though the ice extent fluctuates greatly from year to year.[2] Antarctic sea ice covers a larger area than Arctic ice but is thought to occupy a smaller overall volume, because it's only about half as thick.

The huge Antarctic land ice sheet contains about 90% of the freshwater ice on the earth's surface, and would raise sea levels worldwide by approximately 60 meters (200 feet) were it to melt completely. Whether or not it's melting at all in response to global warming is controversial, as the evidence is mixed. The IPCC's Fifth Assessment Report maintained with high confidence that, from 2005 to 2010, the ice sheet was shedding mass and causing sea levels to rise by 0.41 mm per year (contributing about 24% of the measured rate of 1.7 mm per year between 1900 and 2010).[3] On the other hand, a 2015 NASA study reported that the Antarctic ice sheet was actually gaining rather than losing ice in 2008, and that the ice thickening was making sea levels *fall* by 0.23 mm per year. The study authors found that the ice loss from thinning glaciers in West Antarctica and the Antarctic Peninsula is currently outweighed by new ice formation in East Antarctica resulting from warming-enhanced snowfall.[4] The same authors maintain that the trend has continued until 2018,[5] despite a recent research paper endorsing the IPCC

[1] Crockford, Susan J. 2017. "Baffin Bay and Kane Basin Polar Bears Not 'Declining' Concludes New Report." *Polar Bear Science* blog, 15 February, https://polarbearscience.com/2017/02/15/baffin-bay-and-kane-basin-polar-bears-not-declining-concludes-new-report/. Accessed 12 July 2018.

[2] U.S. Department of Commerce, National Oceanic & Atmospheric Administration. 2018. "State of the Climate: Global Snow and Ice for January 2018," February, https://www.ncdc.noaa.gov/sotc/global-snow/201801. Accessed 12 July 2018.

[3] Vaughan, Comiso, Allison et al., 2013, section 4.4.2.3.

[4] Zwally, H. Jay, Jun Li, John W. Robbins et al. 2015. "Mass Gains of the Antarctic Ice Sheet Exceed Losses." *Journal of Glaciology*, 61, 1019-1036.

[5] Bastasch, Michael. 2018. "Upcoming Research Will Buck the 'Consensus' and Show Antarctica is Still Gaining Ice." *The Daily Caller*, 15 June, https://dailycaller.com/2018/06/15/antarctica-ice-sheets/. Accessed 12 July 2018.

human-caused global warming narrative of decreasing Antarctic ice.[1] While much smaller than Antarctica, the Greenland ice sheet has indeed lost some of its ice by melting since the late 1990s, but not at an alarming rate.

Computer model simulations of sea ice spread that are driven by a combination of human CO_2 emissions and natural variability are in reasonable agreement with observations in the Arctic, but not in the Antarctic. In fact, the fingerprint pattern for the Antarctic is a better match to the observational data when the simulations include only natural causes of global warming.[2] So sea ice isn't a strong line of evidence for anthropogenic climate change. Nor is land ice, if indeed the Antarctic ice sheet is growing rather than shrinking.

Extreme Weather

Yet another line of evidence cited in support of man-made climate change is weather extremes. Since extreme weather events are rare to begin with, however, it's difficult to reliably evaluate any human connection. Even the IPCC concedes that, with the possible exception of record maximum and minimum temperatures, neither heavy downpours, prolonged droughts nor unusually intense tropical cyclones (which include hurricanes and typhoons) can necessarily be ascribed to human influence — in spite of the efforts of politicians and the popular press to proclaim otherwise. Evidence is evidence: to ascertain whether any particular extreme event is becoming more frequent requires observations over an extended period of time, a limitation recognized by a group of climate scientists attempting to assign specific extremes either to natural variability or to human activity.

In the case of temperature, long-term observations are of course available and evidence exists that, at least in some parts of the globe, both warm and cold extremes have become warmer. But whether these changes merely reflect the increase in average daily temperature, or whether they signify some other shift in the pattern of daytime and nighttime temperatures, is uncertain. For instance, the most noticeable change is in warmer minimum temperatures at night. This could simply be due to the upgrade from glass thermometers to electronic sensors discussed earlier, which resulted in an average boost to U.S. minimum temperatures of 0.3° Celsius (0.5° Fahrenheit). And in the

---

[1] The IMBIE team. 2018. "Mass Balance of the Antarctic Ice Sheet from 1992 to 2017." *Nature*, 558, 219-222.
[2] Bindoff, Stott, AchutaRao et al., 2013, figure 10.16.

U.S. also, record high maximums and minimums aren't nearly as numerous today as they were during the Dust Bowl years of the 1930s,[1] despite all the temperature adjustments made since.

As for other extremes, there's a growing realization that anomalous weather events such as hurricanes, floods, droughts and harsh winters have happened for hundreds, if not thousands of years, long before the end of the 19th century when the instrumentation needed to measure extreme weather accurately was first developed. While modern technology such as television and the Internet may have enhanced our awareness of such events, there's no evidence that they have become more frequent. As an example, a recent detailed study of floods across North America and Europe found "no compelling evidence for consistent changes over time" in the occurrence of major floods over the period from 1930 to 2010. The international study authors concluded that the likelihood of a major flooding event in the northern hemisphere during this interval was no different than expected from chance alone, and that the most significant influence on floods was the naturally occurring Atlantic Multidecadal Oscillation.[2] Another study, a survey of extreme weather events since 1900, found surprisingly strong evidence for more frequent weather extremes during the first half of the 20th century than during the second half, which is the period when human emissions of greenhouse gases are claimed to have had the largest effect on our climate.[3]

Extreme weather, therefore, is by far the weakest of the multiple lines of evidence claiming to support the hypothesis of human-induced climate change.

Politics Takes Over Science

Climate is complex. But the essence of science is simple: evidence and logic. As discussed in the previous sections, the evidence for the CO_2 hypothesis of global warming isn't nearly as strong as its advocates claim. Of the multiple lines of so-called evidence for a marked human influence on warming that I've just described, the evidence itself is nonexistent or very weak in two of the six cases: the missing hot spot in the troposphere, and

---

[1] Abatzoglou, John T. and Renaud Barbero. 2014. "Observed and Projected Changes in Absolute Temperature Records Across the Contiguous United States." *Geophysical Research Letters*, 41, 6501-6508.

[2] Hodgkins, Glenn A., Paul H. Whitfield, Donald H. Burn et al. 2017. "Climate-Driven Variability in the Occurrence of Major Floods Across North America and Europe." *Journal of Hydrology*, 552, 704—717.

[3] Kelly, M. J. 2016. "Trends in Extreme Weather Events Since 1900 — An Enduring Conundrum for Wise Policy Advice." *Journal of Geography and Natural Disasters*, 6: 155, 1-7.

extreme weather events. And while the other four cases are unmistakable evidence of increasing temperatures, not a single line of evidence shows a clear fingerprint pattern linking climate change to human activity. The computer climate models needed to make a strong connection between global warming and human emissions of CO_2 fail to capture the recent slowdowns in surface temperature and in ocean heat, don't reproduce observations of sea level rise, and are unable to correctly simulate sea ice extent.

Climate models are the weak link in the chain that proponents of man-made global warming present as unequivocal evidence for a human hand on our climate. They argue that because both CO_2 emissions and global temperatures have gone up since 1850, elevated CO_2 must be the cause of all the warming.[1] This conclusion about cause and effect is faulty reasoning, poor science and no different from the improper use of epidemiology to demonstrate a causal connection in the public health field. With the single exception of smoking and lung cancer, which is 25 to 26 times more prevalent in smokers than in nonsmokers,[2] an epidemiological or observational study can only establish association, not causation, as discussed in Chapter 4. Although it's certain that CO_2 causes *some* warming, there's no empirical evidence that it's 25 or 26 times more likely than natural variability alone to have caused *most* of the observed warming — especially in light of the evidence telling us that, during the period when CO_2 has continuously climbed at an ever increasing rate, the temperature has both risen and fallen for lengthy periods and has barely increased at all since 1998. The implication is that climate models don't adequately simulate sources of natural variability; as discussed, they may overestimate the greenhouse effect anyway.

The underlying reason for these scientific shortcomings and unjustifiable claims can be found in the political sphere, going all the way back to the activities in the late 1970s that led ultimately to the establishment of the IPCC, with its mandate biased toward human-induced climate change. Much more so than the intrusion of politics into nutrition science examined in the previous chapter, the powerful and far-reaching forces of environmental politics have completely subverted climate science in the name of anthropogenic warming.

[1] Oreskes, Naomi. 2007. "The Scientific Consensus on Climate Change: How Do We Know We're Not Wrong?" In Joseph F. C. DiMento and Pamela Doughman, eds., *Climate Change: What it Means for Us, Our Children, and Our Grandchildren*, Cambridge, MA: The MIT Press, 65-99.
[2] U.S. Centers for Disease Control and Prevention. 2014. "Smoking and Cancer," http://www.cdc.gov/tobacco/data_statistics/sgr/50th-anniversary/pdfs/fs_smoking_cancer_508.pdf. Accessed 12 July 2018.

Lest there be any doubt that climate science is being trampled on by politics, it's only necessary to examine the correlation between political views and climate change skepticism. Left-leaning liberals and environmentalists are more than twice as likely to support the conventional wisdom on global warming in countries such as the U.S., the UK and Australia, while the majority of right-leaning conservatives are skeptical about human influence; in the U.S., nearly all Democrats are warmists and nearly all Republicans are skeptics. This type of partisan demarcation in itself should be a warning sign that climate change is much more a political issue than a scientific one.

As discussed earlier in this chapter, the scientific impetus behind the first World Climate Conference in 1979 and the subsequent Villach conferences was the Charney Report. The report concluded that if the CO_2 level continued to increase, there was "no reason to doubt that climate changes will result and no reason to believe that these changes will be negligible"; present-day climate models peg the future warming caused by doubled CO_2 at between 1.5° Celsius (2.7° Fahrenheit) and 4.5° Celsius (8.1° Fahrenheit). The wide range between these projected limits is where the politics enter. If the future temperature gain is only 1.5° Celsius (2.7° Fahrenheit), of which more than half has already occurred, then the world has little to worry about. If it approaches 4.5° Celsius (8.1° Fahrenheit) or even higher, there could be serious consequences, such as widespread flooding of low-lying coastal regions around the world due to higher sea levels.

Politicization of global warming stems from fear about the latter possibility — a fear that was fanned by an ever more strident drumbeat of statements and discussions about the presumed climatic effects of rising greenhouse gas emissions, at gatherings of both climate scientists and politicians in the 1980s and early 1990s. As we know, these activities culminated in the enactment of the UNFCCC, the Kyoto Protocol and the recent Paris Agreement. Fear is often a strong enough force to push science to the sidelines, as Chapters 6 and 7 will also show.

While many climate researchers are well meaning and gather important data, the scheming behind all these political efforts has been greatly aided by exaggeration and manipulation on the part of activist climate scientists. Prominent among these was Hansen, whose testimony at Senate hearings on global warming in 1986 and 1988 was instrumental in making climate change a political issue in the U.S., as noted before. At the 1986 hearing, Hansen presented climate model projections of U.S. temperatures 30 years later that were three times higher than they turned out to be,[1] an exaggeration that set the stage for others who followed, in both the climate science community and the media.

Today, the exaggeration manifests itself most frequently in the setting of a new record for the "hottest year ever." To many in the media and some climate scientists, this is cause for alarm. However, even when the global temperature has almost flatlined and is currently inching upwards at the rate of only a few hundredths of a degree every 10 years, which is a rate of a few thousandths of a degree per year, the establishment of new records is hardly surprising. If the previous record has been set in the last 10 or 20 years, it only requires a high temperature that is several hundredths of a degree above the old record to set a new one. In any case, global warming in itself isn't a sign of human influence on the climate, as we've seen.

Hyping of climate model predictions and new temperature records for political purposes has become part of a deceptive pattern in climate science and is an abuse of science as a whole. As I've already mentioned, distortion of the truth may have even infiltrated government scientific agencies such as NOAA. Another illustration of exaggeration for political reasons is the supposed 97% consensus on man-made global warming among climate scientists which, as discussed earlier, is in reality no more than about 67%. The actual percentage could be even lower, since many climate scientists, particularly younger ones, who are skeptical of a major human contribution to climate change are reluctant to divulge their true opinions in public, for fear of losing their funding or even their jobs. In fact, an increasing number of skeptical research papers are appearing in the climate science literature with a disclaimer acknowledging the conventional wisdom on human-caused warming — so that the authors can get their research grants renewed.

Funding for climate change research can in fact be difficult to obtain for skeptics, especially if the research is on natural sources of rising temperatures such as the sun, rather than the mainstream CO_2 hypothesis. The situation mirrors how critics of the prevailing diet–heart hypothesis in nutrition science were shut out of U.S. government funding from the 1970s onward, after the government had come down on the side of the hypothesis. The U.S. government has invested heavily in climate science in the past, with

[1] See 2 on page 109. Hansen's prediction for the U.S. was based on what he called Scenario A, which assumed a 1.5% yearly increase of the annual increment in the CO_2 level. Actual global temperatures from 1998 onwards have been closest to Scenario C, which was introduced by Hansen in his testimony at the 1988 Senate hearing and would have reduced CO_2 and other greenhouse gas emissions to zero by 2000. However, as this never happened, the various blogs that claim Hansen has been vindicated by the good match between observations and Scenario C are badly mistaken.

annual research funding of approximately $2.5 billion in 2013;[1] the bulk of this went to support research that corroborates man-made global warming. The total amount funneled annually to skeptics by industry (including often maligned fossil-fuel giant ExxonMobil), charitable foundations and think tanks pales in comparison. For climate scientists, it has paid not to bite the hand that feeds.

The subversion of climate science to politics has created a narrative about dangerous anthropogenic climate change, more akin to a religion than science. As we saw in Chapter 2, religious beliefs — in the existence of a God or in reincarnation, for instance — depend on faith, which demands unquestioning acceptance of beliefs that can't be scientifically tested. The climate change narrative requires faith because the evidence itself can't link humans to warming, certainly not dangerous warming, and the only way a link can be made is through computer climate models that fail most tests of their predictive skill. The narrative is fostered by political bodies such as the IPCC, amplified by a compliant press that serves as a global warming cheerleader, and reinforced by duped members of the general public who look on climate change skepticism as heresy. Confirmation bias, which describes the selective evaluation of observational data supporting one's preconceptions or hypothesis, feeds into the narrative as well, just as it bolstered the diet–heart hypothesis for many years before the weight of evidence caused the hypothesis to be discarded.

Another abuse of science is use of the climate change narrative to shut down debate on global warming, usually with the sweeping declaration that the science is settled. Such pronouncements are the very antithesis of science, and have been made from the beginning, for example by then U.S. Senator Gore in 1986 when little empirical evidence had even been gathered. More recently, UK climatologists tried in 2015 to silence the researcher who led the team predicting possible global cooling after 2020, based on a drastic falloff expected in the sun's magnetic field, by demanding that the Royal Astronomical Society withdraw her press release — which the society, fortunately for science, refused to do. Yet another method of closing off debate is misuse of the peer review process in scientific publishing as a weapon against skeptics. Skepticism and open debate have been an intrinsic part of the scientific method since the Greek Golden Age. Suppression of opinion is more the province of a dictatorship or police state than of science,

---

[1] Leggett, Jane A., Richard K. Lattanzio and Emily Bruner. 2013. *Federal Climate Change Funding from FY2008 to FY2014*, Congressional Research Service Report, 13 September, Table 1, https://www.hsdl.org/?view&did=745047. Accessed 12 July 2018.
   This expenditure was for research only. Much more was spent on other areas such as clean energy technologies.

particularly when contrary evidence continues to emerge, such as the warming pause.

Sometimes the climate narrative faithful conspire to suppress the evidence itself. The most famous example of this was the so-called Climategate scandal in 2009, when thousands of embarrassing private emails — thought to have been hacked — between several of the world's top climate scientists were leaked onto the Internet. The misdeeds that came to light included extensive data manipulation; subterfuge to keep temperature data and computer codes from being released to outside researchers who wanted to perform independent analyses; and destruction of records. Towering over all the revelations was the deceptive effort to reconstruct historical temperatures in order to create the now discredited hockey stick. Although apologists have tried to justify this behavior as normal scientific hubris, such tactics are definitely not science.

Likewise, there's no place in science for the *ad hominem* attacks generated by the climate debate, some of which have been just as vicious as those directed at Wegener last century over continental drift. Although the intensity of such attacks seems to have peaked around the time of the contentious COP-15 in Copenhagen in 2009, occasional warmist calls to jail or even execute skeptics en masse are still made,[1] as are death threats against individual scientists on both sides. Even the pejorative use of the term "denier" in the climate change debate is a deliberate political tactic to tar skeptics with immorality, like Holocaust deniers.

But, while intimidation tactics have been successful in muzzling younger skeptics in climate science, older climate scientists and others already well established in their careers are able to dissent from the prevailing wisdom more freely. One prominent U.S. climatologist and lukewarmer skeptic, John Christy, who leads one of two major research groups that analyze NOAA's satellite temperature data, has given testimony several times at Congressional hearings. Another well-known lukewarmer, though an ecologist rather than a climatologist, is Patrick Moore, a former director of Greenpeace who abandoned the organization in 1986 "to develop an environmental policy based on science and logic rather than sensationalism ... and fear."[2] Such

---

[1] Meyer, Warren. 2010. "Why Blowing up Kids Seemed Like a Good Idea," *Forbes*, 7 October, http://www.forbes.com/sites/warrenmeyer/2010/10/07/why-blowing-up-kids-seemed-like-a-good-idea/. Accessed 12 July 2018.

In his article, Meyer refers to several public calls for climate change skeptics to be punished — by jail, execution or, in the case of a tasteless video, blowing up child skeptics and other disbelievers in the theory of anthropogenic global warming.

[2] Moore, Patrick. 2015. From "Should We Celebrate Carbon Dioxide?," the Annual Lecture of the Global Warming Policy Foundation, 15 October, http://www.thegwpf.org/patrick-moore-should-we-celebrate-carbon-dioxide/. Accessed 12 July 2018.

dissenting voices not only may be right about global warming, but are also essential to the integrity of science itself.

Interestingly, it appears that climate scientists who aren't political activists disapprove of the politicizing of their research field. In a 2015 online survey of 651 climate scientists around the world that allowed graduated responses, approximately 61% of the participants were convinced that climate change is or will be the result of human activity, comparable with our previous estimate of the consensus. About half of the respondents agreed that science "should deliver facts, not policies," while 82% thought that scientific debate among scientists should be based on reason and logic, as opposed to emotions and values. Large majorities also thought that in reality, climate change discourse is driven by public or political sentiment, and that scientific ideas have been distorted to bolster political arguments about climate change.[1]

The entwining of climate change with politics, combined with the realization that empirical evidence was lacking, prompted some climate scientists in the 1990s to label the discipline as "postnormal" science. Postnormal science is a topic we'll return to in the final chapter, but is a term coined by philosophers of science specifically to describe the interface between science and society, between science and politics, characterized above all by uncertainty in the science and the consequent disputes that arise in its interpretation. It's closely related to the Precautionary Principle, which is embodied in the UNFCCC and can be stated as:

> When an activity raises threats of harm to human health or the environment, precautionary measures should be taken even if some cause and effect relationships are not fully established scientifically.[2]

As we've seen several times in this chapter, any cause and effect relationship between human CO_2 emissions and global warming is tenuous, so the latter condition for the Precautionary Principle certainly applies. But whether the portion of global warming that is man-made is potentially dangerous remains highly debatable, and world leaders are fooling themselves if they think otherwise. It's questionable whether attempts to shift the world away from

---

[1] Bray, Dennis and Hans von Storch. 2016. *The Bray and von Storch 5th International Survey of Climate Scientists 2015/2016*, HZG Report 2016-2. Geesthacht, Germany: Helmholtz-Zentrum Geesthacht, https://www.academia.edu/26328070/The_Bray_and_von_Storch_5th_International_Survey_of_Climate_Scientists_2015_2016. Accessed 12 July 2018.
  The numbers cited here were obtained by adding the percentage for the strongest response (rating 7) to half of the percentage for the 2nd strongest response (rating 6).
[2] Science and Environmental Health Network. 2018. "Precautionary Principle, Missoula Statement: Conservation Decisions in the Face of Uncertainty," November, http://www.sehn.org/amsci.html. Accessed 12 July 2018.

fossil fuels to other sources of energy can be justified by appealing even to postnormal science; such efforts belong to the realm of politics, not science.

In any case, postnormal science is a far cry from the rich heritage of the scientific method that was conceived by the ancient Greeks and has been refined for hundreds of years since the Scientific Revolution. It's a mistake to think that science has somehow progressed to a more enlightened, postnormal state, in which climate science can be misused for political ends. True science is, and always will be, about empirical evidence and logic, nothing more.

CHAPTER 6. VACCINATION: EXPLOITATION OF FEAR

In this chapter we'll examine an issue even more emotionally charged than evolution or climate change: the question of whether or not we should vaccinate our children, and the risks involved in the vaccination procedure. Despite an abundance of evidence that vaccination has eradicated killer diseases such as smallpox and polio in many countries, and drastically curtailed others such as measles, mumps and pertussis (whooping cough), some diseases are making a comeback because more and more parents are choosing not to vaccinate their children. Some parents are opposed to immunization on religious or philosophical grounds. But most of the vocal anti-vaccine movement, which includes celebrities and popular doctors, insist that vaccines inevitably cause disabling side effects or other diseases, although the bulk of the available scientific data doesn't support such claims.

Underlying the vaccination debate are dueling fears. Fear of dying, or seeing our offspring die, from a deadly disease is what pushes us to vaccinate at all. Fear of side effects, real or imagined, is largely what drives anti-vaccinationists who exploit this fear to broadcast their message, and ignore the logic and evidence that characterize science.

The Vaccination Story

Vaccination originated as a response to smallpox, which was the deadliest infectious disease of all time and only eliminated in 1980. Smallpox killed more humans than the Black Death and the wars of the 20th century combined, felling an estimated 500 million people. Over the centuries its victims embraced both royalty and the poor, as well as more than half the native population of North America, following the introduction of smallpox to the continent by European settlers.

The first historical effort to protect against this highly contagious, painful and often fatal disease is thought to be the process of inoculation, developed by the Chinese and the Turks during the Middle Ages. Inoculation involved deliberately infecting a person with a mild version of smallpox in order to ward off more serious disease — not unlike vaccination today with a live but attenuated vaccine.[1] In the most common method employed for inoculation, infectious material such as smallpox scabs or fluid from pus-filled blisters on a person with smallpox was inserted into superficial scratches made in the patient's skin. After a few days the patient developed the disease, but in a less severe form than naturally acquired smallpox; once the symptoms disappeared after several weeks, the patient had lifetime immunity.

Inoculation was introduced to England from Turkey during a smallpox epidemic in the early 18th century, and was first practiced in the American colonies around the same time. In 1777, then General George Washington ordered that all soldiers in his Continental Army who had not had smallpox be inoculated, during the American Revolutionary War. But, while the procedure became popular in Europe and the U.S., it had the disadvantage that smallpox could be spread to others in the period immediately after inoculation. For this reason, the discovery around 1770 that immunity to smallpox could be induced much more safely by inoculation with cowpox rather than smallpox itself gained attention. Cowpox belongs to the same virus family as smallpox but is much less virulent.

Although he wasn't the first to use cowpox inoculation to create smallpox immunity, English country doctor and scientist Edward Jenner (1749-1823) is credited with publicizing the new process and demonstrating conclusively that it was effective. It was common knowledge at the time that milkmaids were immune to smallpox if they had previously contracted cowpox from the cows they tended. Indeed, most milkmaids in 18th-century England had faces unblemished by smallpox, unlike almost everyone else whose faces bore the scars of the disease they had survived. In 1796, Jenner scraped pus from fresh cowpox lesions on the hands of a milkmaid into a scratch on the arm of an eight-year-old boy. The boy developed a mild fever and a small pus-filled blister that eventually fell off, but didn't become ill. To test that the boy now had immunity to smallpox, about two months later Jenner inoculated

---

[1] Live attenuated vaccines contain live viruses that have been weakened to the point where they can't replicate enough times in the human body to cause disease, but still stimulate an antibody response sufficiently strong to create immunity. Unlike inoculation that made a patient infectious for the first few days after the procedure, live attenuated vaccines aren't known to enable transmission of the virus to others.

him with fluid from a fresh smallpox lesion; the boy wasn't infected. Jenner called the new procedure vaccination, from the Latin word *vacca* for cow, and successfully tested it on dozens of other people including his own son. The older inoculation process, which was renamed variolation,[1] gradually fell out of use and was eventually banned in many countries.

Vaccination for smallpox quickly spread across England and then to Europe and the U.S. Accepted by all social classes, Jenner's vaccine slashed the number of deaths from smallpox in England by half between 1810 and 1820. However, compared with today, 19th-century methods of vaccination were primitive and painful. The procedure began by using a surgical lancet to score the skin of one arm in at least four different places, this being followed by smearing of the vaccine, known then as lymph, into the cuts. In order to maintain the lymph supply, the British government urged public vaccinators to vaccinate from arm to arm. To comply, vaccinators required infants to be brought back eight days after inoculation so that the lymph in their blisters could be transferred directly to the arms of waiting infants. Apart from these indignities and the disfiguring scars often left by the vaccination process, the arm-to-arm method sometimes resulted in serious infection and even the passing on of blood-borne diseases such as tuberculosis and syphilis.[2]

Literally adding insult to injury, the British government made vaccination mandatory. After a group of prominent physicians concerned about epidemic disease had lobbied the government for compulsory vaccination, and an ineffective law had been enacted in 1853, another, much tougher act became law in 1867. The new act spelled out exactly how vaccination would be enforced and by whom. Parents who chose not to vaccinate their children up to 14 years old would first be warned by medical officers if they didn't have a certificate of vaccination, and would then be taken to court if they ignored the warning. Were they unable or unwilling to pay a steep fine plus court costs, their family belongings could be seized and sold at a public auction.

Not surprisingly, this perceived official overreach on smallpox vaccination sparked a major backlash, spawning the world's first anti-vaccine movement, the precursor of modern efforts to resist vaccination against all diseases. Although there are strong social and political overtones to anti-vaccinationism, at root the anti-vaccination movement, then and now, is about fear — fear of the vaccination process itself and of its side effects. In Jenner's time, British anti-vaccine activists proclaimed that his cowpox vaccine contained the "poison of adders, the blood, entrails, and excretions

---

[1] Variola, the medical term for smallpox, is derived from the Latin word *varius*, meaning spotted.

[2] Durbach, Nadja. 2004. *Bodily Matters: The Anti-Vaccination Movement in England, 1853—1907.* Durham, NC: Duke University Press, 3-4.

of bats, toads and suckling whelps." This witches' brew, they believed, could transform a healthy child into a hydra-headed monster or a "hideous foul-skinned cripple." Others were more direct though equally dramatic in their beliefs about the vaccine, propagating fears that vaccination could actually turn people into cows, or at least make them grow horns or snouts, or induce children to moo and graze in the fields. Protestors linked vaccination to the devil and mocked Jenner in public rallies.

Apart from rallies, anti-vaccinationists produced thousands of handbills, posters and pamphlets, and penned letters to newspaper editors, on what they saw as the horrors of vaccination. They refused to accept scientific advances such as the germ theory of disease, introduced in the 1860s, which explained *why* Jenner's vaccine worked: infection with cowpox virus protected against disease caused by the smallpox virus. Anti-vaccine activists also rejected the discovery in the 1880s by Russian zoologist and immunologist Ilya Mechnikov (1845–1916) of phagocytes, a type of germ-devouring white blood cell that plays an important role in the body's immune system and that helps to explain *how* Jenner's vaccine worked.[1] And they ignored the improvements in vaccination safety made since the early 19th century, including the discontinuance of the arm-to-arm method that had on occasion left infants badly infected.

By the 1890s, there were several hundred anti-vaccination leagues in England and Wales, though none in Scotland or Ireland. Unlike today, the most vocal opposition to vaccination in 19th-century Britain came not from the educated upper-middle class, but from the less educated working class and the poor. Yet public auctions of furniture belonging to the poor, who couldn't pay the fines imposed for not vaccinating their children, became a joke when local middle-class activists began to purchase the furniture and return it to the owners. In 1898, the British government withdrew its compulsory vaccination law and replaced it with a law that allowed conscientious objection and abolished all penalties for not vaccinating a child. By the turn of the century, vaccination rates of babies had already plummeted to 20% in some counties and were under 50% in others. The inevitable consequence was that the scourge of smallpox returned, with

---

[1] Offit, Paul A. 2015. *Deadly Choices: How the Anti-Vaccine Movement Threatens Us All*. New York: Basic Books, 117-125.

Phagocytes ingest microbes such as bacteria directly, though not viruses which hide and replicate inside the body's own cells. Virus-infected cells can also be devoured by phagocytes, however, after the cells have been killed by lymphocytes, another type of white blood cell in the immune system.

more than 3,000 deaths occurring from the disease in England and Wales during an epidemic in 1902-03.[1]

Meanwhile in the U.S., the state of Massachusetts had enacted a law in 1809 that gave the government the power to mandate vaccination or quarantine in the event of a smallpox epidemic. Although similar legislation was passed in several other states, especially as a requirement for public school education, it wasn't until a smallpox epidemic in Boston between 1901 and 1903 that any of the state laws were widely enforced. While compulsory vaccination in Boston during the outbreak kept the death toll down to 270 out of 1,600 cases altogether — numbers that undoubtedly would have been much higher if the city hadn't invoked the law — it also provoked a strong reaction from anti-vaccinationists. Activists claimed that compulsory vaccination slaughtered "tens of thousands of innocent children" and was a more important social issue than slavery, drawing the media response that the debate was "a conflict between intelligence and ignorance, civilization and barbarism."[2]

In contrast to British opposition to mandatory vaccination that was characterized mostly by rejection of science, early U.S. anti-vaccinationists saw compulsory intervention as an attack on personal liberty under the Fourteenth Amendment to the Constitution and took the issue to the courts. After several challenges to local laws from organizations such as the Anti-Vaccination League in Brooklyn, the issue came to a head in 1902 during the Boston smallpox epidemic. That was when pastor Henning Jacobson refused to be vaccinated and then to pay a $5 fine after being convicted by a district court for refusing. Jacobson's conviction was upheld by both an appeal court and the state supreme court. Undaunted, he appealed again to the U.S. Supreme Court. His lawyer argued that the vaccination requirement not only infringed Jacobson's individual rights to control his own health, but was also dangerous as it compelled a man "to offer up his body to pollution and filth and disease ... and virtually to say to a sick calf, 'Thou art my savior: in thee do I trust.' " But despite the lawyer's allusion to cowpox vaccine, the Supreme Court ruled in 1905 that the U.S. Constitution doesn't guarantee the freedom to refuse vaccination, stating that "a community has the right to protect itself against an epidemic of disease which threatens the safety of

---

[1] Smallman-Raynor, Matthew and Andrew Cliff. 2012. *Atlas of Epidemic Britain: A Twentieth Century Picture*. Oxford: Oxford University Press, 38.

[2] Gostin, Lawrence O. 2005. "*Jacobson v Massachusetts* at 100 Years: Police Power and Civil Liberties in Tension." *American Journal of Public Health*, 95, 576-581.

its members." This landmark ruling endorsed both science and the ethical principle that public health trumps individual rights to one's own body.[1]

Following Jenner's great stride in pioneering smallpox vaccination at the end of the 18th century, it wasn't until nearly a hundred years later that the vaccination technique was extended to other infectious diseases, when Louis Pasteur (1822-95) created a vaccine for rabies. Vaccines for yet other diseases were developed during the 20th century, which is when the scientific field of immunology bloomed.

In 1921, there were 206,000 cases of diphtheria[2] and 15,500 deaths in the U.S. alone.[3] U.S. cases of pertussis, like diphtheria also a bacterial disease, numbered 265,000 in 1934, of which 7,500 died. And the viral disease polio at its U.S. peak in 1952 permanently paralyzed 21,000 children, killing 3,000 of them.[4] By 1990, thanks to the introduction of vaccines for these and other diseases such as measles, mumps and rubella (German measles), they had all been either completely eliminated or their incidence reduced by more than 90%. Today, unfortunately, some of these old diseases are on the rise again because so many parents are opting out of vaccinating their children.

Vaccine Safety on Trial

Just as with Jenner's smallpox vaccine, safety issues, both real and imaginary, have arisen with vaccines that protect against other infectious diseases. Vaccination, like many medications and surgical procedures, is not entirely free of side effects. But most are mild and transient such as soreness or swelling at the injection site, or a temporary fever or rash; permanent, debilitating side effects are extremely rare. The risk of a serious adverse reaction to vaccination is typically about one in a million, but ranges from zero (with no known serious side effects) up to one in a thousand, depending on the vaccine.[5] However, these numbers are minuscule compared with the risk of contracting the disease itself in an unvaccinated community during

---

[1] Parmet, Wendy E., Richard A. Goodman and Amy Farber. 2005. "Individual Rights Versus the Public's Health — 100 Years After *Jacobson* v. *Massachusetts*." *New England Journal of Medicine*, 352, 652-654.

[2] Diphtheria is a bacterial infection that produces a thick covering on the back of the throat, which can suffocate the victim.

[3] U.S. Centers for Disease Control and Prevention. 2016. "Diphtheria: About Diphtheria," https://www.cdc.gov/diphtheria/about/. Accessed 12 July 2018.

[4] Roush, Sandra W., Trudy V. Murphy et al. 2007. "Historical Comparisons of Morbidity and Mortality for Vaccine-Preventable Diseases in the United States." *Journal of the American Medical Association*, 298, 2155-2163.

[5] World Health Organization. 2018. "Global Vaccine Safety: WHO Vaccine Reaction Rates Information Sheets," http://www.who.int/vaccine_safety/initiative/tools/vaccinfosheets/en/. Accessed 12 July 2018.

an epidemic. As we'll see later in this chapter, unvaccinated children are up to nine times or more likely to catch chicken pox, whooping cough or pneumonia than vaccinated children exposed to the same bacteria or viruses. Vaccine-preventable diseases not only kill a significant number of those afflicted, but can have devastating aftereffects such as paralysis, blindness, deafness or mental retardation in those who survive.

The vaccine historically linked to the most side effects is the DTP vaccine, a combination that immunizes simultaneously against diphtheria, tetanus and pertussis. The original DTP vaccine, first available in 1948, was something of a chemical soup as it contained over 3,000 pertussis proteins, in addition to a single diphtheria and a single tetanus protein — the proteins or antigens being the active agents that stimulate the production of disease-fighting antibodies once inside the human body. In contrast, the combination measles-mumps-rubella (MMR) vaccine contains only 24 proteins, the polio vaccine only 15. It was the large number of proteins in the early DTP vaccine that caused so many more reactions, albeit only temporary, than other vaccines. But the DTP vaccine has also been unjustifiably blamed for a litany of permanent side effects including seizures, paralysis, epilepsy, other neurological disorders, brain damage and mental retardation — even Sudden Infant Death Syndrome (SIDS). Although there are several possible short-term side effects from the vaccine, there are no known long-term effects.[1]

The claim that DTP vaccine permanently harmed or killed children first came to widespread public notice in England in 1973, when pediatric neurologist John Wilson presented a paper at a meeting of the Royal Society of Medicine. With the measured demeanor of a bishop delivering a sermon, Wilson related the stories of 36 children who had been his patients at the Hospital for Sick Children in London because of neurological illness that he linked to pertussis immunization. Supposedly, some 22 of the children had suffered epileptic seizures and were mentally disabled; one child had vomited for four days after her second DTP vaccination, become blind and then died of a seizure six months later.[2] Wilson's presentation created a furor, both in the medical community and the media. Unsurprisingly, the DTP immunization rate in Britain slumped from 77% in 1974 to 30% in 1978,[3] causing cases of whooping cough to soar from 12,000 to 67,000 annually over

[1] Barlow, William E., Robert L. Davis, John W. Glasser et al. 2001. "The Risk of Seizures After Receipt of Whole-Cell Pertussis or Measles, Mumps, and Rubella Vaccine." *New England Journal of Medicine*, 345, 656-661.
[2] Kulenkampff, M., J. S. Schwartzman and J. Wilson. 1974. "Neurological Complications of Pertussis Inoculation." *Archives of Disease in Childhood*, 49, 46-49.
[3] U.S. Centers for Disease Control and Prevention. 1982. "International Notes Pertussis — England and Wales." *Morbidity and Mortality Weekly Report* (MMWR), 31, 629-631, https://www.cdc.gov/mmwr/preview/mmwrhtml/00001197.htm. Accessed 12 July 2018.

that period. At the end of the decade, 36 children died in one of the worst epidemics in modern history.[1]

Wilson's startling assertions were reinforced by the results of a major British study conducted between 1976 and 1979 on neurological illnesses thought to be caused by DTP vaccine. The study claimed to find a statistically significant risk of one in one hundred thousand children experiencing a serious neurological disorder within seven days of a DTP injection, especially in the first 72 hours; the disorders included encephalitis[2] and seizures. But the study, carried out by a team of London doctors led by David Miller, also found that most children recovered and "showed no evidence of residual damage one year later."[3]

Despite this qualification by the authors, the Miller study along with publicity over Wilson's earlier paper triggered an avalanche of lawsuits against vaccine manufacturers and doctors, in both the UK and U.S. Between 1981 and 1986, the annual number of lawsuits filed in the U.S. over assumed side effects from DTP vaccine jumped from 3 to 255. Damages sought and actual awards climbed too, with jury awards and out-of-court settlements often being in the million-dollar range.[4] As a result, vaccine prices escalated and pharmaceutical companies began to abandon the vaccine business; by early 1986 only two suppliers of DTP vaccine remained. Finally, Congress took action by establishing the National Vaccine Injury Compensation Program later that year. The legislation set up a government compensation procedure for children who had experienced seizures and brain damage allegedly caused by DTP vaccine, specifying award amounts and legal fees. Not only did the program give parents the option of an administrative settlement instead of a prolonged and expensive court case, but it also protected vaccine manufacturers from excessive litigation.[5]

In the less litigious UK, alarm over DTP vaccination led to only three trials. The plaintiff in the first trial, a mentally retarded nine-year-old boy, lost his case. The second trial ended without a verdict when the plaintiff's mother lied, claiming that her son had suffered seizures that started seven hours after a DTP injection, whereas hospital records showed the boy's first seizure actually occurred five months later. The third trial, which

---

[1] Deer, Brian. 1998. "The Vanishing Victims." *The Sunday Times Magazine*, 1 November, http://briandeer.com/dtp-dpt-vaccine-1.htm. Accessed 12 July 2018.

[2] Inflammation of the brain.

[3] Miller, D. L., E. M. Ross, R. Alderslade et al. 1981. "Pertussis Immunisation and Serious Acute Neurological Illness in Children." *British Medical Journal*, 282, 1595-1599.

[4] Hinman, A. R. 1986. "DTP Vaccine Litigation." *American Journal of Diseases of Children*, 140, 528-530.

[5] Smith, Martin H. 1988. "National Childhood Vaccine Injury Compensation Act." *Pediatrics*, 82, 264-269.

could have paved the way for a class-action lawsuit involving two hundred other children, involved a 17-year old girl with severe autism, a condition that her mother claimed was a direct result of three doses of DTP vaccine administered to her daughter as an infant.[1] Among the expert witnesses at the lengthy trial in 1987-88 were John Wilson, the neurologist whose paper had sparked the vaccine scare in the 1970s, and David Miller, coauthor of the later study purporting to quantify the risk of neurological disorders caused by DTP immunization.

In his judgment, which at more than 100,000 words was as long as this book, Judge Murray Stuart-Smith was scathing toward both these witnesses. He chided Wilson for including in his paper two children who had received just DT vaccine (without the pertussis component), not DTP, and for forgetting this fact in his trial testimony "because it was inconvenient." But the judge saved his strongest excoriation for Wilson's admission to the court that of his original 36 cases supposedly linking neurological illness to DTP vaccination, he now stood by only 12, all of which involved seizures that could have had alternative explanations. Most damning of all, two of those 12 were twin girls who had subsequently died from a rare genetic disorder but, it was later discovered, had never been given the DTP vaccine.

In Miller's case, Stuart-Smith faulted important details in the study reported in 1981. Even though the Miller team had recorded that 12 previously normal children developed a neurological illness which persisted for at least a year within seven days of being immunized with DTP vaccine, three of the children were in fact normal both before and after vaccination. The neurological symptoms in three others were the result of viral infections, not vaccination, and one child had Reye's Syndrome, a problem later discovered to be caused by aspirin, not vaccines. The judge was also bothered that the Miller study had been published prematurely, before all the collected data had been analyzed. Publication of the complete data in 1993, well after the trial, revealed that indeed the original study had drawn invalid conclusions.[2]

Stuart-Smith's trial verdict was unequivocal, stating that the "plaintiff had failed to establish ... that pertussis vaccine as administered in the UK ... could cause permanent brain damage in young children."[3] The verdict, together with an identical decision in a landmark case in Canada the same year and the introduction of the National Vaccine Injury Compensation

---

[1] Dyer, Clare. 1987. "Whooping Cough Vaccine on Trial Again." *British Medical Journal*, 295, 1053-1054.
[2] Miller, David, Nicola Madge, Judith Diamond et al. 1993. "Pertussis Immunisation and Serious Acute Neurological Illnesses in Children." *BMJ*, 307, 1171-1176.
[3] Offit, 2015, 22-23, 32-40.

Program in the U.S., helped stem the flow of litigation against DTP vaccine. By 1996, the annual number of lawsuits had drastically fallen.

But despite the falloff in legal action, there was still little scientific evidence to dispel the then popular belief that vaccination against pertussis could lead to brain damage. Because the Miller team's survey had been of limited extent, several larger-scale investigations were conducted during the 1980s in the UK and the U.S. — at the very same time that lawsuits over the vaccine's effects were proliferating. A comprehensive seven-year study of two separate groups of approximately 134,000 children in England, one of which had been given DTP vaccine but the other group DT vaccine alone, showed "no convincing evidence that DTP vaccine caused major neurological damage."[1] A U.S. study of 218,000 children likewise revealed no statistically significant risk of acute neurological illness within seven days of vaccination,[2] in direct conflict with the Miller results. Several other studies came to the same conclusion.

Now faced with overwhelming evidence that DTP vaccine didn't cause permanent side effects after all, the U.S. government in 1995 removed residual seizure disorders (epilepsy) from the list of compensable illnesses for the vaccine. In 2001, the issue was finally laid to rest scientifically by an extensive study of seizures following vaccination in almost 700,000 U.S. children. While the study found a significantly increased risk of febrile seizures on the day of DTP vaccination, there was no elevated risk thereafter and no apparent long-term, adverse consequences; febrile seizures are triggered by fever but don't result in permanent harm. Pertussis vaccination became safer yet in the mid-1990s, when newer DTP vaccines were licensed containing only two to five proteins instead of the earlier 3,000 or more.

A second vaccine linked to serious side effects is the MMR vaccine, especially the measles component. The most serious adverse effect is encephalitis, occurring in approximately one of every three million doses of the vaccine. However, this needs to be compared with the much, much higher risk of about one in a thousand for falling prey to encephalitis after contracting measles itself, a viral disease that has killed more children than any other disease in history.[3] But the MMR vaccine has been falsely tied to other side effects, ever since U.S. anti-vaccinationist and activist Barbara

---

[1] Pollock, T. M. and Jean Morris. 1983. "A 7-year Survey of Disorders Attributed to Vaccination in North West Thames Region." *The Lancet*, 321, 753-757.

[2] Gale, James L., Purushottam B. Thapa, Steven G. F. Wassilak et al. 1994. "Risk of Serious Acute Neurological Illness After Immunization with Diphtheria-Tetanus-Pertussis Vaccine: A Population-Based Case-Control Study." *Journal of the American Medical Association*, 271, 37-41.

[3] Biss, Eula. 2014. *On Immunity: An Inoculation*. Minneapolis, MN: Graywolf Press, 35, 107.

Loe Fisher first suggested in the 1980s that vaccines might cause autism spectrum disorders. The MMR vaccine came under the spotlight in 1998 when British gastroenterologist Andrew Wakefield connected autism with MMR vaccination, in a paper reminiscent of Wilson's linking neurological disorders to DTP vaccine 25 years earlier. We'll return to this topic in the next section (on the modern anti-vaccination movement), as the story involves scientific fraud, a phenomenon unfortunately becoming increasingly common today, as well as vaccine side effects.

Another vaccine wrongfully blamed for side effects it didn't cause is the Hib vaccine. Hib is a horrible bacterial infection, particularly for young children: its aftereffects include meningitis[1], which can be deadly and can leave survivors deaf or mentally retarded; pneumonia; and epiglottitis.[2] Nevertheless, the astounding claim of anti-vaccinationists wasn't that the Hib vaccine, first licensed in 1985, was ineffective in combating such devastating side effects, but that the vaccine caused the debilitating, lifelong disease of diabetes. Although the claim was promoted by a doctor on U.S. national television in 1998, it turned out to be based solely on a small study of Hib vaccination on children in Finland. But other researchers were unable to replicate the Finnish study, including a team that compared the risk of succumbing to diabetes in 21,000 American children who had been vaccinated against Hib with 21,000 who had not: the risk of diabetes was the same in both groups. Replication of results, as we saw in Chapter 1, is a central step in the scientific method. What is established, however, is that vaccination for Hib has prevented approximately 25,000 cases of meningitis, pneumonia and bloodstream infections each year in the U.S.

Other vaccines linked by anti-vaccinationists to an array of side effects, including SIDS, diabetes, multiple sclerosis, paralysis and even death, are those that protect against hepatitis B, pneumococcal infections and human papillomavirus (HPV). However, just as with DTP and Hib vaccines, no scientific evidence has emerged to support any of the claims, despite numerous studies that investigated each of the supposed side effects. SIDS, it was discovered in the 1990s, was most likely to result from a child sleeping face down rather than on its back, not from being vaccinated.[3] Again, the reality is that vaccination against hepatitis B, pneumococcus and HPV

---

[1] Inflammation of the brain and spinal cord.
[2] Epiglottitis is a swelling of the epiglottis in the throat, the flap that normally closes in order to cover the windpipe while swallowing. The swelling blocks air flow to the lungs even when the epiglottis is open.
[3] Offit, Paul A. and Charlotte A. Moser. 2011. *Vaccines and Your Child: Separating Fact from Fiction*. New York: Columbia University Press, 59-64.

has each year prevented many tens of thousands of serious infections with potential long-term aftereffects.

One vaccine that did have adverse side effects was the oral polio vaccine, before it was taken off the market around the turn of this century. The oral vaccine, invented by U.S. medical researcher Albert Sabin (1906-93) as a more appealing form of vaccination than injection, was given the green light by the U.S. Public Health Service in 1961. Sabin's vaccine soon replaced the earlier injectable polio vaccine, introduced by U.S. virologist Jonas Salk (1914-95) in the 1950s and tested in a trial on almost two million children.[1] Polio was at the time one of the most feared infectious diseases, because it struck without warning and often crippled its victims through paralysis, if not killing them. Paralyzed patients often had to use an unsightly breathing machine known as an iron lung for lengthy periods.

The Salk and Sabin vaccines were quite different. Salk's was a killed-virus vaccine, one of the first of its type: the vaccine is produced by growing live virus and then inactivating it chemically.[2] Sabin's oral vaccine, on the other hand, was a live-virus vaccine: live virus is weakened below the level where it can normally cause disease;[1] the MMR vaccine is a modern example. Unable to attract support for another large-scale trial like Salk's in the U.S., Sabin persuaded the Soviet Union to conduct a field trial of his vaccine in 1957. But despite its extensive adoption around the world from the 1960s onward, and the subsequent gigantic drop in polio cases, it was unfortunately found that the live Sabin vaccine could occasionally mutate into a strong enough form to cause paralytic polio — a problem that occurred in about one in 2.5 million children each year, and resulted in death for some of the very unlucky victims. While this risk is no higher than the risk of serious side effects from any vaccine, the U.S. decided in 1999 to return to the Salk inactivated vaccine, a move that was soon followed by several other countries.

In the same year, another vaccine was removed from the market in the U.S. This was a newly developed vaccine for protection against a common intestinal virus called rotavirus, a disease responsible for the deaths of thousands of children every day in the developing world and about 60 per year in the U.S. But almost as soon as the vaccine was introduced in the U.S., a large increase was noted in children suffering from an intestinal blockage known as intussusception, which is a serious medical emergency with sometimes fatal complications; all the children had recently received

---

[1] Chemical Heritage Foundation. 2015. "Jonas Salk and Albert Bruce Sabin," https://www. chemheritage.org/historical-profile/jonas-salk-and-albert-bruce-sabin. Accessed 12 July 2018.

[2] Usually with formaldehyde, which kills viruses and bacteria, and is also used to preserve tissue specimens.

the rotavirus vaccine. An investigation found that children who had been vaccinated were up to 25 times more likely to develop intussusception than those who had not, though the risk fell off the longer the time after vaccination and the higher the number of vaccine doses.[1] But as a result of the investigation, the vaccine was withdrawn from the market; a safer vaccine did not emerge for another seven years.

While this episode with rotavirus vaccine may have added to the fears of anti-vaccinationists concerned about vaccine safety, it should in fact have reassured them because the vaccine was taken off the market so promptly. The U.S. Centers for Disease Control (CDC) acted quickly in recognizing the problem and conducting several extensive scientific studies. In any case, the episode is far from a regular occurrence: it was the first time a U.S. vaccine had been discontinued over a safety issue since the so-called Cutter incident in 1955. That incident involved killed-virus polio vaccine manufactured by Cutter Laboratories that had been recalled because it had not been fully inactivated, and therefore contained live polio virus, a mistake that resulted in many children being paralyzed and 10 deaths.[2]

The Anti-Vaccination Movement Revives

As discussed earlier in the chapter, the anti-vaccine movement had its roots in the 19th century when smallpox vaccination became mandatory, both in Britain and in certain American states. But as more and more people were vaccinated and so-called herd immunity built up, the prevalence of smallpox diminished, at least in industrialized countries. By 1930, anti-vaccination organizations had largely disbanded — only to regroup again in the U.S. during the 1970s and 1980s, following the introduction of DPT, MMR and polio vaccines in the intervening years, and a subsequent increase in the number of state laws mandating vaccination for children in public and private schools. In the UK, however, compulsory vaccination was finally abandoned in 1946.

Many state mandatory vaccination laws ("no shots, no school") in the U.S. were the result of measles outbreaks in the 1960s and 1970s. Even though the introduction of a measles vaccine in 1963 had lowered reported U.S. measles cases from hundreds of thousands to a few tens of thousands

---

[1] U.S. Centers for Disease Control and Prevention. 2011. "Vaccines & Immunizations: Rotavirus Vaccine (RotaShield®) and Intussusception," https://www.cdc.gov/vaccines/vpd-vac/rotavirus/vac-rotashield-historical.htm. Accessed 12 July 2018.

[2] U.S. Centers for Disease Control and Prevention. 2018. "Vaccine Safety: Cutter Incident — 1955," https://www.cdc.gov/vaccinesafety/concerns/concerns-history.html. Accessed 12 July 2018.

per year by 1968, this number crept back up to 75,000 in 1971.[1] But as states updated their earlier smallpox vaccination laws to include immunization against measles and other diseases, the measles caseload came down again. By 1980, all 50 U.S. states had instituted school immunization requirements. Following the official recommendation in 1989 that all children receive not one but two doses of measles vaccine, usually as the MMR combination, the incidence of measles dropped off to only 100 cases per year by 1998.[2]

School mandates have been extremely effective in reducing the prevalence of all major infectious diseases in the U.S. Nevertheless, some parents have opposed the mandates for religious reasons and have regarded the denial of education for unvaccinated children as an imposition on their religious freedom, in much the same way that Jacobson at the beginning of the century had regarded compulsory vaccination as infringing on his constitutional rights. So, like Jacobson, later 20th-century parents took their grievances to the courts — except that they claimed protection under the First Amendment, which guarantees freedom of religion, instead of the Fourteenth Amendment governing civil rights. But none of the First Amendment lawsuits succeeded. In one of the cases, the judge ruled that: "The right to practice religion freely does not include liberty to expose the community or the child to communicable disease...,"[3] echoing the Jacobson Fourteenth Amendment ruling in 1905. In another case, the judge declared that the health of school children outweighed the religious beliefs of a parent.[4]

However, the playing field changed drastically when New York State in 1966 opened the door to religious exemptions by enacting a new law allowing parents to opt out of a polio vaccination requirement for school entry. Originally limited to parents who were members of a recognized religious organization, the law was later broadened to include anyone who held "genuine and sincere religious beliefs" contrary to vaccination.[5] The law resulted from intensive lobbying by Christian Scientists, one of the most influential religious groups in the U.S., who believe that prayer can overcome disease. Nonetheless, public health records show that several measles outbreaks occurred among Christian Scientist high-school students

---

[1] U.S. Centers for Disease Control and Prevention. 1994. "Summary of Notifiable Diseases, United States 1993." *Morbidity and Mortality Weekly Report* (MMWR), 42, 38, https://www.cdc.gov/Mmwr/PDF/wk/mm4253.pdf. Accessed 12 July 2018.

[2] Dennehy, Penelope H. 2001. "Active Immunization in the United States: Developments over the Past Decade." *Clinical Microbiology Reviews*, 14, 872-908.

[3] Supreme Court of Arkansas. 1965. Wright v. DeWitt School District No. 1 of Arkansas County, 385 S.W.2d 644.

[4] Supreme Court of Mississippi. 1979. Brown v. Stone, 378 So.2d 218.

[5] Kraus, Caroline L. 2001. "Religious Exemptions — Applicability to Vegetarian Beliefs." *Hofstra Law Review*, 30, 197-224.

in the St. Louis, Missouri area between 1985 and 1994, in one of which three students died.[1] In Connecticut in 1972, a polio outbreak at a Christian Science boarding school left 11 children with varying degrees of paralysis.[2]

Religious exemptions paved the way for philosophical exemptions, based on strongly held personal, moral or other nonreligious beliefs. Because such beliefs are more difficult to define in legal terms than religious objections to vaccination, fewer U.S. states allow philosophical exemptions than grant religious exemptions. While 46 states allowed religious exemptions in 2016, only 19 offered philosophical exemptions. All 50 states, however, provide exemptions on medical grounds, such as children who can't be vaccinated at all owing to chemotherapy for cancer, immunosuppressive therapy for a transplant or steroid therapy for asthma.

But religious and genuine philosophical exemptions constitute only a small fraction of overall objections to vaccination. By far the greatest number of objections come from those who are fearful of side effects. We've already seen how the DTP vaccine and vaccines that induce immunity against Hib and other infectious diseases have been wrongfully blamed for an extensive range of side effects. A more recent example is the association of the MMR vaccine with autism, the neurodevelopmental disorder that can be so worrying to parents. The autism episode generated worldwide publicity and led to thousands of court cases in a special U.S. Vaccine Court set up as part of the National Vaccine Injury Compensation Program, far more cases than the hundreds of lawsuits filed in the U.S. and UK over presumed side effects from DTP vaccine.

It was Andrew Wakefield's 1998 paper[3] that set the MMR vaccine–autism debate ablaze. Wakefield claimed that eight out of 12 children had developed symptoms of autism following MMR vaccination, and that all eight had intestinal inflammation. He linked the inflammation to the vaccine and proposed that the children's upset intestines leaked harmful proteins into the bloodstream, where they were carried to the brain, causing autism. With the flames fanned by anti-vaccinationists, hundreds of thousands of parents who had previously adhered dutifully to immunization schedules panicked and began refusing MMR vaccine. Needless to say, outbreaks

---

[1] U.S. Centers for Disease Control and Prevention. 1994. "Outbreak of Measles Among Christian Science Students — Missouri and Illinois, 1994." *Morbidity and Mortality Weekly Report* (MMWR), 43, 463-465, https://www.cdc.gov/mmwr/preview/ mmwrhtml/00031788.htm. Accessed 12 July 2018.

[2] Foote, F. M., G. Kraus, M. D. Andrews and J. C. Hart. 1973. "Polio Outbreak in a Private School." *Connecticut Medicine*, 37, 643-644.

[3] Wakefield, A. J., S. H. Murch, A. Anthony et al. 1998. "Ileal-Lymphoid-Nodular Hyperplasia, Non-Specific Colitis, and Pervasive Developmental Disorder in Children." *The Lancet*, 351, 637-641.

of measles subsequently occurred all over the world, just as public health officials thought the two-dose vaccination recommendation had conquered the disease. However, Wakefield's paper was slowly discredited, piece by piece, over the next 12 years. A number of studies showed that MMR vaccine doesn't cause intestinal inflammation; brain-damaging proteins could not be found in the blood of vaccinated children; and multiple epidemiological studies revealed no link between MMR vaccination and autism.[1]

As Wakefield's credibility was being undermined by the lack of supporting evidence from colleagues, other irregularities in his work were coming to light. In 2004, the *Sunday Times* of London exposed a previously unknown payment of £55,000 that he had received from a legal aid fund in 1996, specifically to investigate a possible link between MMR vaccine and autism in five of the children in his study, whose parents were suing vaccine manufacturers over the claimed link.[2] Wakefield was later paid approximately £390,000 more from the same legal aid fund. But he had kept this conflict of interest a secret and had not declared it to *The Lancet* which published his paper, nor to his coauthors. The month after the *Sunday Times* 2004 article, 10 of the 12 coauthors withdrew their endorsement of the paper's claims. In 2007, England's General Medical Council began investigating Wakefield for medical misconduct: the charges included subjecting children to unnecessary spinal taps and intestinal biopsies, and paying them £5 for blood samples while attending his son's birthday party. The Council ruled in 2010 that he had shown "callous disregard" for the suffering of children in his study, failed to obtain the proper ethics approval for the research, and was dishonest and irresponsible in publishing his results.[3]

Just one week after the Medical Council's verdict, *The Lancet* formally retracted Wakefield's 1998 paper. This was an extremely serious step in itself for the journal to take, as retraction implies a belief by the journal editors that the study was falsified or misrepresented. In 2011, the editors went one step further by declaring the 1998 paper fraudulent, the most serious charge possible, citing clear evidence of falsification of the data on autism and the MMR vaccine.[4] The editors drew on both the Medical Council's findings and a very thorough probe of the individual children's cases by investigative

[1] Offit, 2015, 92-96.
[2] Deer, Brian. 2004. "Revealed: MMR Research Scandal." *Sunday Times*, 22 February, http://briandeer.com/mmr/lancet-deer-1.htm. Accessed 12 July 2018.
[3] Press Association. 2010. "MMR Doctor 'Failed to Act in Interests of Children' ." *The Guardian*, 28 January, https://www.theguardian.com/science/2010/jan/28/mmr-doctor-fail-children-gmc. Accessed 12 July 2018.
[4] Godlee, Fiona, Jane Smith and Harvey Marcovitch. 2011. "Wakefield's Article Linking MMR Vaccine and Autism was Fraudulent." *BMJ*, 2011; 342: c7452.

journalist Brian Deer[1] — who found that not one of the 12 cases reported in the 1998 *Lancet* paper was untainted by "misrepresentation or undisclosed alteration." Wakefield, by then working in America, lost his medical license in 2010; so did his leading coauthor.

The final word on MMR vaccine and autism came not from the General Medical Council or *The Lancet*, but from the U.S. Vaccine Court. Unlike earlier court cases in which seizures and neurological disorders had been falsely blamed on the DTP vaccine, the cases that pinned autism on the MMR vaccine weren't focused on the vaccine administered alone, but rather in conjunction with other, non-MMR vaccines containing a mercury-based preservative known as thimerosal. Thimerosal has been used to prevent bacterial contamination of vaccines since the 1930s. Since 2001, however, its use has been discontinued in all U.S. vaccines except some flu shots, as a result of Food and Drug Administration (FDA) concerns over ingestion of mercury, especially by infants.[2] As noted previously, multiple studies have shown no link between autism and MMR vaccination on its own,[3] and in fact the incidence of autism continued to rise after thimerosal was all but abolished in 2001.[4]

The Vaccine Court began hearing evidence on MMR vaccine, thimerosal and autism in 2007. To handle the enormous caseload of more than 5,600 cases, the court assigned three special masters to hear just three test cases on each of two theories: that autism was caused by a combination of MMR vaccine and thimerosal-containing vaccines, or that it was caused by thimerosal-containing vaccines alone. Despite the limited number of cases heard, the final evidentiary record for each theory totaled tens of thousands of pages, including court transcripts, post-hearing briefs, medical records and journal papers.[5]

---

[1] Deer, Brian. 2011. "Secrets of the MMR Scare: How the Case Against the MMR Vaccine was Fixed." *BMJ*, 2011; 342: c5347.

[2] U.S. Department of Health and Human Services, Food and Drug Administration. 2018. "Vaccine Safety & Availability: Thimerosal in Vaccines," http://www.fda.gov/BiologicsBloodVaccines/SafetyAvailability/VaccineSafety/UCM096228. Accessed 12 July 2018.

    Thimerosal contains approximately 50% ethylmercury by weight. Ethylmercury is more readily excreted from the body than methylmercury, which is a neurotoxin that accumulates in certain types of fish.

[3] See, for example, Parker, Sarah K., Benjamin Schwartz, James Todd and Larry K. Pickering. 2004. "Thimerosal-Containing Vaccines and Autistic Spectrum Disorder: A Critical Review of Published Original Data." *Pediatrics*, 114, 793-804.

[4] Schechter, Robert and Judith K. Grether. 2008. "Continuing Increases in Autism Reported to California's Developmental Services System: Mercury in Retrograde." *Archives of General Psychiatry*, 65, 19-24.

[5] U.S. Court of Federal Claims, Office of Special Masters. 2010. "The Autism Proceedings," http://www.uscfc.uscourts.gov/sites/default/files/vaccine_files/autism.background.2010.pdf. Accessed 12 July 2018.

The special masters' verdicts in the three test cases on the first theory, issued on February 12, 2009, unanimously rejected the contention that MMR vaccine together with thimerosal-containing vaccines could cause autism.[1] Throughout the proceedings, the special masters made it clear they would be guided only by scientific evidence, not by sentiment — of which there was an abundance, as one would expect in hearing the poignant stories of autistic children. In her analysis of the evidence, one of the masters stated:

> Sadly, the petitioners in this litigation have been the victims of bad science, conducted to support litigation rather than to advance medical and scientific understanding of autism spectrum disorder. The evidence in support of petitioners' causal theory is weak, contradictory, and unpersuasive.[2]

Likewise, the verdicts in the three test cases on the second theory, presented a year later on March 12, 2010, ruled that the plaintiffs had failed to establish that autism could be caused or even aggravated by thimerosal-containing vaccines on their own. The special masters called the thimerosal theory "scientifically unsupportable," and chastised doctors and researchers who "peddled hope, not opinions grounded in science and medicine."[3] Autism aside, however, thimerosal has not been a component of MMR and most other vaccines since 2001 as we saw above, even though prominent anti-vaccinationists such as Robert Kennedy Jr. have continued to insist otherwise.[4]

And despite the Vaccine Court's deliberations in the U.S. and *The Lancet*'s accusation of fraud against Wakefield in the UK, anti-vaccinationists, many passionate in their beliefs, continue to link the MMR vaccine to autism. In 2016, Wakefield had the effrontery to direct a documentary, "Vaxxed,"[5] alleging that the U.S. CDC covered up contrary data in a 2004 study that

---

[1] U.S. Court of Federal Claims, Office of Special Masters. 2009. Cedillo v. Secretary of Health and Human Services, 98-916V; Hazlehurst v. Secretary of Health and Human Services, 03-654V; Snyder v. Secretary of Health and Human Services, 01-162V, http://www.uscfc.uscourts.gov/autism-decisions-and-background-information. Accessed 12 July 2018.

[2] U.S. Court of Federal Claims, Office of Special Masters. 2009. Snyder v. Secretary of Health and Human Services, 01-162V 208, http://www.uscfc.uscourts.gov/sites/default/files/vaccine_files/Vowell.Snyder.pdf. Accessed 12 July 2018.

[3] U.S. Court of Federal Claims, Office of Special Masters. 2010. Dwyer v. Secretary of Health and Human Services, 03-1202V; King v. Secretary of Health and Human Services, 03-584V; Mead v. Secretary of Health and Human Services, 03-215V, http://www.uscfc.uscourts.gov/autism-decisions-and-background-information. Accessed 12 July 2018.

[4] Kennedy, Robert F. Jr. 2005. "Deadly Immunity." *Rolling Stone*, 20 June, https://web.archive.org/web/20060422012127/http://www.rollingstone.com/politics/story/7395411/deadly_immunity/. Accessed 12 July 2018.

[5] Cinema Libre Studio. 2016. "Vaxxed: From Cover-Up to Catastrophe," http://vaxxedthemovie.com/. Accessed 12 July 2018.

found MMR vaccine doesn't increase the risk of autism — the same finding as that of the Vaccine Court hearings and numerous epidemiological studies. However, the 2014 research paper forming the basis of Wakefield's audacious claim was subsequently retracted,[1] just like his own paper, and the CDC continues to stand by its 2004 conclusion rejecting any link between autism and the MMR vaccine.[2]

According to the CDC, the prevalence of autism spectrum disorder has more than doubled since 2000, the condition afflicting 1 in 59 U.S. children in 2014.[3] Much of this apparently increased incidence likely results from greater awareness of the disorder and improved detection of what previously was undiagnosed or misdiagnosed. Nevertheless, diagnosis of the condition can be devastating and highly stressful for the desperate parents of an autistic child, who naturally tend to grasp for explanations and are often perfectly willing to believe the hype about vaccination. Currently, the causes of autism remain unknown, although a number of risk factors have been identified: certain genetic conditions have been implicated, and it's thought that exposure during pregnancy to toxic chemicals such as pesticides, or to bacterial or viral infections, plays a role.[4] Because autism is such a complex problem, it may require artificial intelligence, with its ability to sift vast amounts of data, to explain where the disorder comes from.

In the U.S., the modern anti-vaccine movement has become highly visible through its promotion by celebrities. These include former Playboy playmate Jenny McCarthy (whose son is autistic), actor Jim Carrey, political talk show host Bill Maher, television and radio host Larry King and many others. Unfortunately for science, the celebrities tend to reinforce the movement's dependence on anecdote and hearsay, rather than solid scientific evidence.

For example, a popular anti-vaccination claim is that it's dangerous for babies to be given up to five vaccines at once, because their nascent immune systems can be overwhelmed. In a 1999 survey, 25% of parents felt their

---

[1] Hooker, Brian S. 2014. "Retraction Note: Measles-Mumps-Rubella Vaccination Timing and Autism Among Young African American Boys: A Reanalysis of CDC Data." *Translational Neurodegeneration*, 3, 22.

[2] U.S. Centers for Disease Control and Prevention. 2015. "Vaccine Safety: CDC Statement Regarding 2004 Pediatrics Article, 'Age at First Measles-Mumps-Rubella Vaccination in Children with Autism and School-Matched Control Subjects: A Population-Based Study in Metropolitan Atlanta' ," https://www.cdc.gov/vaccinesafety/concerns/autism/cdc2004pediatrics.html. Accessed 12 July 2018.

[3] U.S. Centers for Disease Control and Prevention. 2018. "Autism Spectrum Disorder (ASD): Data & Statistics," https://www.cdc.gov/ncbddd/autism/data.html. Accessed 12 July 2018.

[4] CNBC. 2017. "Scientists Narrow Down the Startling Risk Factors that Can Cause Autism," 7 September, https://www.cnbc.com/2017/09/07/scientists-reveal-autism-risk-factors.html. Accessed 12 July 2018.

children's immune systems could be damaged by excessive vaccinations.[1] Although there's absolutely no medical evidence for such a claim, more research into the possible effects of multiple vaccinations at the same time could help to reassure worried parents. However, one scientific study of the response of the infant immune system to multiple vaccinations has shown that an infant could theoretically tolerate as many as 10,000 vaccines at once, though this is a purely hypothetical number since the total number of vaccines recommended for infants in the U.S. is fewer than 20. The same study has pointed out that the total number of proteins — the antigens in vaccines that stimulate an immune response and are the cause of short-term side effects — in all vaccines recommended today is far fewer than the 200 or so proteins in the single smallpox vaccine administered at the beginning of the 20th century.[2]

Nevertheless, the "Too many, too soon!" rallying cry of anti-vaccine activists has achieved a small degree of credibility through its endorsement by qualified doctors. Among these is California pediatrician Robert Sears, known as Dr. Bob, who advocates an alternative U.S. vaccine schedule that spaces out infant vaccinations over a longer period than the CDC recommended timetable. But there's no evidence that a drawn-out schedule is of any benefit to children. In fact, Sears admitted in a 2015 interview that there were no published, peer-reviewed papers to support his notion of "antigenic overload";[3] several peer-reviewed papers show the contention to be false.[4]

Another of Sears' claims about vaccination is that vaccines have a high rate of serious side effects, based on his review of data from the U.S. Vaccine Adverse Event Reporting System (VAERS). Created in 1990 to serve as an early warning system for potential safety issues in vaccines, VAERS was instrumental in identifying intussusception as a major side effect of the newly introduced rotavirus vaccine in 1999, a problem I discussed in the previous section. VAERS defines a serious side effect as one that results in "permanent disability, hospitalization, life-threatening illness or death," severe reactions

[1] Mooney, Chris. 2009. "Why Does the Vaccine/Autism Controversy Live on?" *Discover*, 6 May, http://discovermagazine.com/2009/jun/06-why-does-vaccine-autism-controversy-live-on. Accessed 12 July 2018.

[2] Offit, Paul A., Jessica Quarles, Michael A. Gerber et al. 2002. "Addressing Parents' Concerns: Do Multiple Vaccines Overwhelm or Weaken the Infant's Immune System?" *Pediatrics*, 109, 124-129.

[3] MSNBC video. 2015. "Challenging Vaccine Delayers," All In with Chris Hayes, 12 February, http://www.msnbc.com/all-in/watch/challenging-vaccine-delayers-398406723692. Accessed 12 July 2018.

[4] Poland, Gregory A. and Robert M. Jacobson. 2012. "The Clinician's Guide to the Anti-Vaccinationists' Galaxy." *Human Immunology*, 73, 859-866.

that occur in 10% to 15% of cases reported to VAERS. Together with VAERS statistics, such a percentage suggests an actual number of several thousand severe reactions per year, which is the number cited by Sears in another 2015 interview.[1] However, there's no way of knowing whether these reactions come from a vaccine or not, since VAERS encourages the reporting of *any* serious health problem that occurs after vaccination, even when it's uncertain or unlikely that the vaccine caused the problem. The CDC itself states that such problems are "rarely caused by the vaccine."[2]

A better gauge of the actual number of serious adverse reactions to vaccination is the number of petitions submitted to the U.S. Vaccine Court that are awarded compensation. Because of the financial element, petitions to the court are more likely to reflect actual side effects from vaccines than reports made to VAERS, although the court emphasizes that a compensation award doesn't necessarily imply that a vaccine caused the alleged injury. According to the CDC, the Vaccine Court compensated 3,723 individuals from 2006 to 2016, during which time over 3.1 billion doses of covered vaccines were distributed in the U.S.[3] This means that one person was compensated for almost every one million doses of vaccine, consistent with the one in a million estimated risk of a serious adverse reaction to vaccination discussed earlier.

A further assertion by Sears and many anti-vaccinationists is that vaccines contain harmful or toxic chemicals such as aluminum, formaldehyde and thimerosal. As we've seen, thimerosal is now used only in some influenza vaccines. Formaldehyde is used in killed-virus vaccines to inactivate the live virus, but the amount of formaldehyde is a tiny fraction of that found in many foods, including those fed to babies such as pureed bananas or pears.

Aluminum salts are employed as adjuvants to enhance the immune system response to a vaccine, thus reducing the number of doses needed. But what is important about these and other vaccine ingredients is that they're present only in minuscule quantities, well below the threshold for toxicity, and many of the ingredients are already present in the body. Aluminum, for instance, is found in our food supply, in drinking water and even in the air we breathe, as well as being ingested by babies in breast milk or infant formula. Because it has been suggested that aluminum could be linked to certain neurological

[1] CNN Tonight. 2015. "104 Cases of Measles Across 14 States," 2 February, http://www.cnn.com/TRANSCRIPTS/1502/02/cnnt.01.html. Accessed 12 July 2018.

[2] U.S. Centers for Disease Control and Prevention. 2017. "Vaccine Safety: Vaccine Adverse Event Reporting System (VAERS)," https://www.cdc.gov/vaccinesafety/ensuringsafety/monitoring/vaers/. Accessed 12 July 2018.

[3] U.S. Department of Health and Human Services, Health Resources and Services Administration. 2018. "Vaccine Injury Compensation Data," https://www.hrsa.gov/vaccine-compensation/data/index.html. Accessed 12 July 2018.

disorders such as Alzheimer's Disease, anti-vaccinationists maintain that injected aluminum rapidly enters the bloodstream and thereby accumulates in the brain, causing neurological damage — reminiscent of Wakefield's fraudulent claim about autism. But if this were so, ingested aluminum would have the same effect on the brain, since a portion of the much larger amount of ingested (compared to injected) aluminum also enters the bloodstream before being eliminated from the body via the kidneys. Needless to say, there's no medical evidence to support the injected aluminum hypothesis.

And allegations that some vaccines contain phantom ingredients such as ether, antifreeze or human blood, modern versions of the belief in Jenner's day that his cowpox vaccine included products derived from snakes and bats, are simply fictitious. Science is ill-served by such ignorance and fear.

Fear Negates Science

Both supporters of vaccination and many anti-vaccinationists agree on one thing: that vaccination wards off infectious diseases and saves lives. That's a basic scientific observation made over and over again, on many occasions and for many diseases, as we've seen a number of times in this chapter. As a present-day example, a pediatrician and epidemiologist who teamed up to investigate the vaccination history of hundreds of thousands of children in Colorado found that unvaccinated children were about 23 times more likely to contract whooping cough, nine times more likely to catch chicken pox and 6.5 times more likely to be hospitalized with pneumonia or pneumococcal disease than vaccinated children from the same communities.[1]

Where the two sides diverge is on the scientific case for the safety of vaccines. There's no question that vaccination *can* have serious side effects, which are indeed something to fear, but the important scientific issue is relative risk. As I've discussed, the typical risk of a serious adverse reaction to a vaccine shot is around one in a million. Should your own child be that one in a million, then naturally it would be a traumatic experience to see your previously healthy pride and joy suddenly become incapacitated or even die. But such a horrible thought has to be balanced by the reality that leaving your child unvaccinated could expose him or her to the ravages of a disease with equally or even more disastrous consequences.

Unvaccinated children exposed to someone infected with measles, for instance, have a 90% chance of contracting the disease,[2] and one of 1,000

[1] Daley, Matthew F. and Jason M. Glanz. 2011. "Straight Talk about Vaccination." *Scientific American*, 305, 32-34.
[2] Hassink, Sandra G. 2015. "American Academy of Pediatrics President Urges Parents to Vaccinate Their Children Against Measles," 2 February, https://www.aap.org/en-us/

children who get the measles will develop encephalitis, an often debilitating aftereffect that can lead to seizures and mental retardation. That's a risk of about one in 1,100 of experiencing a serious aftereffect for a child who comes down with measles, compared with the much lower risk of one in one million following vaccination. While it could be argued that the one in 1,100 risk of a serious aftereffect from measles would be much lower if the child never came in contact with a measles-infected person, the chances of avoiding such contact are slim during measles epidemics — which, as noted several times already, aren't rare even in this age of mass immunization. In claiming that vaccines are unsafe and should be avoided, "anti-vaxxers" cast logic aside and ignore the science telling us that it's a thousand times *more* risky for a child to forgo vaccination and suffer the consequences of catching a potentially deadly disease, than to be vaccinated against it.

In the event of a genuine safety problem with a vaccine, we've seen how the U.S. at least has an effective watchdog in the form of the CDC. Although the NIH failed in its oversight of Salk polio vaccine production in 1955, resulting in the infamous Cutter incident mentioned earlier, the only potential problems since then have been nipped in the bud by the CDC. The Sabin live-virus polio vaccine, which occasionally became strong enough through mutation to cause paralytic polio instead of protecting against it, and the original rotavirus vaccine, which put vaccinated children at very high risk for developing intestinal blockage, were both withdrawn from the U.S. market in 1999 through CDC action. Counterparts of the CDC in other countries such as Japan, Canada, Australia and the UK have also withdrawn specific vaccines when abnormally high rates of side effects were reported. However, vaccine withdrawals are rare events today, simply because the vaccine industry is closely regulated and a new vaccine is only allowed on the market in the first place after successfully undergoing an extensive evaluation in three separate phases of clinical trials, which are the gold standard in medical and pharmaceutical testing.

The actual statistical risk of a serious adverse reaction to vaccination, apart from being much lower than the risk of being afflicted with the disease itself, is also a great deal less than the danger involved in common surgical procedures. For example, the risk of dying from bleeding following tonsillectomy, a surgery frequently performed on children and young adults, is as high as one in 12,000 in Israel,[1] a country whose health care system ranks

---

about-the-aap/aap-press-room/Pages/American-Academy-of-Pediatrics-President-Urges-Parents-to-Vaccinate-Their-Children-Against-Measles.aspx. Accessed 12 July 2018.

[1] Cohen, D. and M. Dor. 2008. "Morbidity and Mortality of Post-Tonsillectomy Bleeding: Analysis of Cases." *Journal of Laryngology and Otology*, 122, 88-92.

among the world's best. As a rather different comparison, the odds of a car occupant dying in an accident on U.S. roads were about one in 48,000 in 2013.[1]

Nonetheless, in spite of rational scientific assessments of risk, anti-vaccinationists driven by fear choose to focus not on relative risk, but on the one in a million adverse reactions to a vaccine. Neil Miller, who isn't a doctor, has recently summarized 400 scientific research papers that document the details of adverse reactions to various vaccines, including some containing thimerosal.[2] A slightly smaller list was compiled by the U.S. Institute of Medicine in 2012.[3] While Miller's list consists of largely valid observations of negative and occasionally fatal side effects from vaccination, his 400 cases have to be weighed against the vast number of vaccinations administered without any reported side effects. In the U.S., hundreds of millions of vaccine doses are administered every year; as there are no harmful reactions to the overwhelming majority of these, no medical research papers are written about them. Many hundreds of thousands of papers, however, have been published about vaccination and its beneficial effects, compared with only perhaps several thousand papers about its occasional adverse outcomes.

Some anti-vaccinationists, when accused of exposing their children to unnecessary risk by refusing or delaying immunization because of overblown fears about vaccine safety, justify their stance by appealing to herd immunity. Just as sheer numbers protect a herd of animals from predators, herd immunity protects a community from an infectious disease once enough people become immune to that disease through vaccination. Immunization of more and more members of the community makes it more and more difficult for viruses and bacteria to spread; when a sufficiently large number of people have been vaccinated, infections can no longer circulate at all. The threshold for herd immunity, which is the minimum percentage of the population that needs to be vaccinated for the herd effect to take hold, depends on the contagiousness of the disease, the effectiveness of the vaccine and other factors. For highly contagious measles, herd immunity requires up to 94% of the populace to be immunized, while for polio the percentage ranges up to a slightly lower 86%.[4]

---

[1] Insurance Information Institute. 2018. "Mortality Risk." http://www.iii.org/fact-statistic/mortality-risk. Accessed 12 July 2018.

[2] Miller, Neil Z. 2016. *Miller's Review of Critical Vaccine Studies: 400 Important Scientific Papers Summarized for Parents and Researchers*. Santa Fe, NM: New Atlantean Press.

[3] U.S. Institute of Medicine, Board on Population Health and Public Health Practice. 2012. *Adverse Effects of Vaccines: Evidence and Causality*; Kathleen Stratton, Andrew Ford, Erin Rusch and Ellen Wright Clayton (eds.). Washington, DC: National Academies Press.

[4] Willingham, Emily and Laura Helft. 2014. "What is Herd Immunity?" *NOVA*, 5 September, http://www.pbs.org/wgbh/nova/body/herd-immunity.html. Accessed 12 July 2018.

That the threshold is lower than 100% enables anti-vaxxers to hide their children in the herd. By not vaccinating their offspring but choosing to live in a vaccinated community, anti-vaccinationists avoid the low but real risk of their children experiencing serious side effects from vaccines, while at the same time not exposing them to anyone infected with the disease — at least not in their own community. It would seem that herd immunity enables anti-vaccine advocates to get the best of both worlds: no vaccines and no disease.

But deciding not to vaccinate and hiding in the herd are personal choices that take advantage of the sacrifice of others and amount to free riding, because these personal decisions affect everyone. Just as U.S. courts have ruled over the years, societal ethics dictate that public health should override personal freedom. In any case, hiding in the herd is irrational. If too many people choose not to vaccinate against a particular contagious disease, the percentage vaccinated will fall below the threshold and herd immunity will break down. Once this happens, the disease will be able to sweep rapidly through the population, including both anti-vaccinationist holdouts and the more conscientious vaccinated members of the community. But the people that will suffer the most are those who can't be vaccinated at all, such as children who are allergic to vaccines or have weakened immune systems from undergoing chemotherapy for leukemia or other cancers, and the elderly on immunosuppressive therapy for rheumatic diseases. These more vulnerable groups depend for their protection on those around them being vaccinated; the level of community protection increases with each successive vaccination. It's the immunologically weak who have no choice but to depend on herd immunity, and who are especially susceptible to the choices made by anti-vaccinationists.

An issue closely related to herd immunity is religious and philosophical exemptions from mandatory vaccination. As we saw in the previous section, most U.S. states currently allow religious exemptions and about 40% allow philosophical exemptions to state laws mandating vaccination for schoolchildren. Although the laws originated in the 19th century, many of them were a reaction to recurrent outbreaks of measles in the 1960s and 1970s. But almost as soon as school mandates were in place in all 50 states by 1980, the number of nonmedical exemptions began to grow. This growth in exemptions was fueled by two main factors: complacency over the greatly reduced prevalence of vaccine-preventable diseases, and an increase in public fears about adverse health effects from vaccines. At the same time, the accompanying rise of the anti-vaccination movement led to the lowering of herd immunity for many infectious diseases, causing ongoing outbreaks in the U.S. and elsewhere of whooping cough, measles and mumps.

In fact, an examination in 2006 of nonmedical exemptions in the U.S. showed that the highest incidence of diseases such as whooping cough, both in exempted children and in their communities, occurred in the states that were most generous in granting exemptions.[1] Again and again, it's been demonstrated that states with lower exemption rates enjoy lower levels of infectious disease. This scientific evidence of the efficacy of vaccination has prompted a number of states to take a tougher stand on allowing exemptions, and even to abolish nonmedical exemptions altogether.

The pushback against exemptions has received a further boost from repeated outbreaks of measles. Following a record number of 667 cases of measles from 23 outbreaks reported in the U.S. in 2014[2] — a substantial escalation from the 100 cases documented in 1998 — a 2015 outbreak that started in Disneyland, California reinvigorated the debate over mandatory vaccination. The outbreak, which sickened 131 Californians and spread to other states as well as Canada and Mexico, was blamed on too few of those who fell ill being immunized against measles.[3] A study at the time concluded that the MMR vaccination rate among visitors and workers exposed to measles during the outbreak was well below the threshold for herd immunity, possibly as low as 50%.[4] A similar phenomenon was observed in Minnesota in 2017, when a measles outbreak which sickened more than 40 Somali American children was blamed on an MMR vaccination rate in the Somali community that had fallen from 92% in 2004 to only 42% in 2014.[5] In response to the mass exposure to this deadly disease in California, legislators introduced a bill to repeal the state's religious and personal belief

[1] Omer, S. B., W. K. Pan, N. A. Halsey et al. 2006. "Nonmedical Exemptions to School Immunization Requirements: Secular Trends and Association of State Policies with Pertussis Incidence." *Journal of the American Medical Association*, 296, 1757-1763.

[2] U.S. Centers for Disease Control and Prevention. 2018. "Measles (Rubeola): Measles Cases and Outbreaks," https://www.cdc.gov/measles/cases-outbreaks.html. Accessed 12 July 2018.

[3] NBC News. 2015. "Measles Outbreak Traced to Disneyland is Declared Over," 17 April, http://www.nbcnews.com/storyline/measles-outbreak/measles-outbreak-traced-disneyland-declared-over-n343686. Accessed 12 July 2018.

[4] Majumder, Maimuna S., Emily L. Cohn, Sumiko R. Mekaru et al. 2015. "Substandard Vaccination Compliance and the 2015 Measles Outbreak". *JAMA Pediatrics*, 169, 494-495.

[5] Sun, Lena H. 2017. "Anti-Vaccine Activists Spark a State's Worst Measles Outbreak in Decades." *Washington Post*, 5 May, https://www.washingtonpost.com/national/health-science/anti-vaccine-activists-spark-a-states-worst-measles-outbreak-in-decades/2017/05/04/a1fac952-2f39-11e7-9dec-764dc781686f_story.html?utm_term=.daf39cdf3cfc. Accessed 12 July 2018.

exemptions for vaccination of schoolchildren.[1] After a bitter and protracted debate, during which anti-vaccinationists packed legislative hearings and bombarded lawmakers with phone calls and petitions, the bill became law in 2015: under the new law, only medical exemptions are still allowed.

While science may have gained the upper hand in the California debate, the passage of the bill was nevertheless accompanied by virulent *ad hominem* attacks against the legislation's principal author, pediatrician and state senator Richard Pan — vitriol just as fierce as that directed at Wegener during the continental drift debate discussed in Chapter 2, or that aimed at skeptics of man-made global warming discussed in Chapter 5. Posts on Pan's Facebook page called for him to be "eradicated" or hung by a noose, in addition to other death threats.[2] As I've said before, such vehement animosity in scientific discussion is an abuse of science itself. California anti-vaccinationists may even have recognized this, since a subsequent effort to recall Pan failed for lack of support.

The anti-vaccine movement's lack of solid scientific underpinnings is perhaps best exemplified by the scientific fraud involved in Wakefield's falsification of data, allegedly linking the MMR vaccine to autism, and to a lesser extent by the false claim that DTP vaccine could cause permanent brain damage in young children. As we've seen, it took a court of law to establish how unscientific the evidence for both of these contentions was. The use of litigation to resolve scientific issues reveals a parallel between anti-vaccinationists and the anti-evolutionists of Chapter 3. Anti-evolutionists resorted to the legal system, for example in the Scopes Monkey Trial, to argue unsuccessfully that creationism should be taught in place of or alongside evolution in public schools; creationism has its roots in religious beliefs, not science. Anti-vaccine advocates have attempted to use lawsuits and the courtroom to argue, again unsuccessfully, that their objections to vaccination are scientific and that vaccines are harmful; but anti-vaccinationists are motivated by fear rather than science.

As the next chapter will also show, fear can be as much of a threat to science and the scientific method as politics is in the debates over dietary fat and climate change.

---

[1] Pan, Richard, California State Senate. 2015. "Senate Bill 277 Introduced to End California's Vaccine Exemption Loophole," 19 February, http://sd06.senate.ca.gov/news/2015-02-19-senate-bill-277-introduced-end-california%E2%80%99s-vaccine-exemption-loophole. Accessed 12 July 2018.
[2] *Sacramento Bee.* 2015. "From Death Threats to Holocaust Warning, California Vaccine Bill an Extraordinary Fight," 30 June, http://www.sacbee.com/news/politics-government/capitol-alert/article25909216.html. Accessed 12 July 2018.

CHAPTER 7. GMO FOODS: FEAR OF A FRANKENSTEIN

In this final example of the attack on modern science, we'll look at the debate over foods containing genetically modified organisms (GMOs), which have now been on the market for more than 20 years. Genetically engineered crops were originally developed as a more scientific alternative to traditional methods of plant breeding, in the hope of boosting yields and nutritional content sufficiently to keep pace with the needs of the world's burgeoning population. While this goal remains elusive, almost 40% of a wary U.S. public believes that GMO foods are worse for our health than non-GMO foods, despite the lack of any scientific evidence that GMOs have ever caused harm to a human.[1] In the UK, where cultivation of GMO crops is currently banned, 40% of the population is opposed to any promotion of GMOs by the government.[2]

Driving the anti-GMO movement is the fear that tinkering with a food plant's genes could create "Frankenfoods," evocative of the monster created by the fictional mad scientist Frankenstein — plants containing unknown viruses or toxins, or capable of wreaking havoc with the environment. Fears over food safety and the ecology of the planet, coupled with widespread suspicion of the motives of the agribusiness companies who supply the

---

[1] Hefferon, Meg and Monica Anderson. 2016. "Younger Generations Stand Out in Their Beliefs About Organic, GM Foods," *Pew Research Center*, 7 December, http://www. pewresearch.org/fact-tank/2016/12/07/younger-generations-stand-out-in-their-beliefs-about-organic-gm-foods/. Accessed 12 July 2018.
[2] Knapton, Sarah. 2015. "Genetically Modified Crops Could be Planted in England This Year." *The Telegraph*, 13 January, http://www.telegraph.co.uk/news/earth/agriculture/geneticmodification/11343502/Genetically-modified-crops-could-be-planted-in-England-this-year.html. Accessed 12 July 2018.

seeds, fertilizers, herbicides and pesticides for GMO crops, have brought environmental and agricultural politics into the debate. The result has been that the underlying science is frequently distorted or ignored.

The GMO Foods Story

Although GMO crops and foods are a relatively recent phenomenon, human modification of nature's food supply dates all the way back to the dawn of agriculture about 11,000 years ago. That's when hunter-gatherers began to settle down in permanent villages and to cultivate wild grains such as wheat, barley and rice.

To improve the yields of their domesticated crops, early farmers replanted the seeds of only the largest and healthiest looking plants; other characteristics artificially selected likely included taste and endurance of harsh weather conditions. Apart from making farming more productive, this selective breeding created plant varieties over the centuries that today bear no resemblance whatsoever to their wild ancestors. Modern corn on the cob, for example, with its multiple tidy rows of plump kernels, is a far cry from the spindly Mexican grass, with no more than 12 scrawny kernels each enclosed in a stony case, that it evolved from.[1]

It wasn't until the 18th and 19th centuries that plant breeding became a science rather than an art, following the discovery of pollen as the agent of plant sexuality and the precise role it played in fertilization. This led to experiments in artificial hybridization or crossbreeding that for the first time combined different varieties of a plant in the breeding process, as opposed to the centuries-old selective breeding involving a single plant variety. The early crossbreeder pollinated by hand, using simple tools such as tweezers or a scalpel to emasculate one set of plants and a paintbrush to collect male pollen from the plants to be crossbred. The pollen was then dusted on the feminized plants. One of the most well-known examples is the Russet Burbank (Idaho) potato, the popular oblong staple developed in the 1870s by prolific American plant breeder Luther Burbank (1849–1926), and now used to make french fries eaten all over the world. The same method was soon extended to the hybridization of different but related species such as different types of vegetable or fruit, for instance potatoes crossed with tomatoes or peaches with plums.[2] Crossbreeding also occurs naturally, with

---

[1] Carroll, Sean B. 2010. "Tracking the Ancestry of Corn Back 9,000 Years." *New York Times*, 24 May, http://www.nytimes.com/2010/05/25/science/25creature.html. Accessed 12 July 2018.

[2] Fedoroff, Nina V. and Nancy Marie Brown. 2004. *Mendel in the Kitchen: A Scientist's View of Genetically Modified Foods*. Washington, DC: Joseph Henry Press, 51-54.

pollen being transported by insects or the wind; most pollination of current commercial crops is carried out by honeybees.

The early 20th century saw major improvements to artificial hybridization techniques. In 1917, U.S. plant breeder and geneticist Donald F. Jones (1890–1963) invented the double-cross method of hybrid seed production, a breakthrough that paved the way for commercialization of high-yield hybrid corn. The double-cross method utilizes four inbred[1] parent lines. Crossing of only two inbred lines resulted in yields that were much too low for mass production, though the single-cross hybrids were more robust plants with bigger ears than their parents. Hybrids continued to improve, revolutionizing the growing of corn: the average yield per acre in the U.S. tripled from 1929 to 1969, as the hybrid percentage of all corn grown shot up from zero to more than 99%.[2]

But the biggest contribution of hybridization, aside from its practical successes in food cultivation, was to the scientific understanding of plant breeding through the pioneering experiments of the Austrian monk and botanist Mendel in the 19th century. As mentioned in Chapter 3, Mendel's work on genetic inheritance went largely unnoticed at the time, even though it provided an explanation for the mechanism of natural selection in evolution that Darwin put forward in his 1859 *Origin of Species*. The methodical Mendel, a farmer's son, took crossbreeding to a new level in his modest monastery garden by initiating a seven-year study of the humble pea, crossing round peas with wrinkled ones, peas from yellow pods with those from green pods, tall plants with diminutive ones.[3] He carefully charted his results, noting how many round or wrinkled, yellow or green, tall or short peas appeared in the next generation. Although his mathematical approach was unfamiliar to biologists, causing his discovery to be ignored for the next 35 years, it was Mendel who first proposed that characteristics such as the height, color and shape of organisms depend on what we now call genes — discrete units that transmit dominant and recessive traits from parents to offspring.

---

[1] Inbreeding in plants is a result of self-pollination from the same plant, repeated for several generations until the progeny are genetically identical to each other and to the parent. [Plant & Soil Sciences eLibrary. 2018. "Corn Breeding: Lessons from the Past — Inbreeding, Hybrid Vigor, and Hybrid Corn," http://passel.unl.edu/pages/informationmodule.php?idinformationmodule=1075412493&topicorder=9&maxto=12. Accessed 12 July 2018.]

[2] Mangelsdorf, Paul C., National Academy of Sciences. 1975. "Donald Forsha Jones: April 16, 1890—June 19, 1963." In *Biographical Memoirs, Volume XLVI*, Washington, DC: National Academies Press, chap. 5.

[3] In peas (and corn), both male stamens and female pistils are parts of the same plant. Given enough time, such plants normally pollinate themselves.

Mendel's new laws countered the contemporary theory in Darwin's day that inheritance hinged on blending, rather than transmission, of parental traits.

Both traditional crossbreeding and modern genetic engineering involve gene transfer; selective breeding also involves genetic change, but through mutation rather than human intervention. Both crossbreeding and genetic engineering can result in unexpected and unintended genetic effects, despite claims to the contrary by advocates for both GMO and non-GMO foods. In conventional crossbreeding, thousands of unknown genes whose functions are also unknown are transferred to the host plant, along with the intended genes. In mutational crossbreeding via radiation or chemicals, hundreds or thousands of random mutations are induced in each mutated line. Unexpected genetic changes occur all the time in the wild too.

Some scientists consider genetic engineering to be more precise than conventional breeding because only known and well understood genes are transferred. But alien genes that land in unpredictable sections of the host plant's genetic code or genome[1] could, under certain conditions, create new allergens or toxins in the edible parts of the plant. However, traditional plant breeding methods can also produce potentially hazardous foods. At least two commercial varieties of crossbred potato have been withdrawn from the market because of elevated, toxic levels of normally innocuous chemical compounds. Even the potato varieties still on the market contain natural toxins in trace, harmless amounts, as do celery, carrots, mustard and a host of other foods.[2]

There are several genetic engineering methods for splicing an alien gene into a food plant's genome. One frequently used technique consists of "cutting and pasting" DNA from the genome that contains the gene or genes to be transferred, the DNA snipping being done with molecular scissors known as restriction enzymes, of which more than 3,000 are known. To transfer the cut DNA fragments to the host plant, the DNA is first inserted inside a bacterial cell called *Agrobacterium*, a common bacterium in soil. The bacterium invades plants through wounds in the root or stem and normally causes a tumor-like growth by injecting its bacterial DNA into the plant's

---

[1] A genome is an organism's complete set of genetic instructions, including all its genes. The genome is stored in long molecules of DNA called chromosomes: chromosomes are to the genome as chapters are to a book and, in this analogy, genes are the pages or stories within each chapter.
[2] U.S. Institute of Medicine and National Research Council. 2004. *Safety of Genetically Engineered Foods: Approaches to Assessing Unintended Health Effects*. Washington, DC: National Academies Press, chap. 3, https://www.nap.edu/read/10977/chapter/5. Accessed 12 July 2018.

genome.[1] The trick involved in this genetic engineering technique is to have the alien genes inserted into the bacterium *replace* its tumor-inducing genes, so that a tumor no longer grows but the host plant exhibits the new genetic trait instead. In practice, the engineered *Agrobacterium* cells, sometimes called a bacterial "truck," are cultured with tiny pieces of the plant in a suitable liquid medium, allowed to infect them, and the whole plant then regenerated from the infected pieces. A rather different technique is to propel the foreign genes directly into the host plant's tissue by means of a gene "gun." This involves shooting tiny gold or tungsten particles coated with the alien DNA through the intact cell walls of the plant. Less common techniques include electric shock and chemical treatments.

The principal drawback to genetic engineering, which has the advantage over traditional crossbreeding of being able to transfer very specific, known genes to the transgenic plant, is that the location of the transferred genes in the host genome can't be precisely pinpointed. As already mentioned, this can lead to unexpected results in genetically engineered plants — just as in crossbred plants, although the surprises in hybridization arise from the hitchhiker genes that tag along with the intended genes inserted into the host genome, and not from uncertainty in the gene location. Unintended effects in both breeding methods include the failure of successive generations of the plant to reproduce the desired trait, and excessive production of unwanted or toxic proteins. What all this means for genetic engineering in practice is that new GMO foods need to be exhaustively tested before being allowed to enter the marketplace. But because GMO crops are regulated more extensively than conventional crossbred plants, potentially hazardous traits are more likely to be detected at an early, developmental stage in genetically engineered crops.

A possible concern in genetic engineering is the existence of "jumping genes" [2] — genes capable of hopscotching about, either within the same genome, or even from one species of plant or organism to an unrelated species, by either copy-and-paste or cut-and-paste mechanisms. An example from another field is antibiotic resistance in bacteria, where a gene imparting resistance to an antibiotic such as penicillin can jump to another bacterial species, making that species penicillin resistant as well.[3] Jumping genes were

[1] The genetic vehicle that transports the alien genes is a short, often circular segment of bacterial DNA, usually distinct from the bacterium's chromosomes, called a plasmid.

[2] In genetics, jumping genes are called transposons. The jumping process is one of several methods of horizontal gene transfer, so called to distinguish it from vertical gene transfer from parents to offspring that takes place in evolution.

[3] Sharps, Eric. 2013. "I Eat DNA — There's Just No Case Against GMO Foods." *FourSquare Partners Magazine* blog, 8 October, http://foursquarepartners.blogspot.com/2013/10/i-eat-dna.html#!/2013/10/i-eat-dna.html. Accessed 12 July 2018.

discovered by geneticist Barbara McClintock (1902-92) in the late 1940s, as an outcome of her painstaking studies of corn (maize), and her observations about strange speckles sometimes seen on decorative Indian corn. She was finally awarded a Nobel Prize for her findings in 1983. Many more jumping genes were uncovered in the years after McClintock's discovery, and in fact they're present in nearly all organisms, including humans. Nonetheless, despite the prevalence of these maverick residents in the genome, more often than not they're defective or silent, and remain out of the action because it's to the organism's evolutionary advantage.

One of the most vocal recent critics of genetic engineering was geneticist and biotechnology activist Mae-Wan Ho (1941–2016). In the 1990s, Ho took issue with a so-called gene transfer promoter known as CaMV 35S.[1] Promoters are an essential part of the gene package or cassette that accompanies gene transfer in plants, playing an important role in determining where and when the new gene will be switched on in the transgenic plant. The CaMV 35S is an especially energetic promoter, driving genes at an unusually frenzied pace, for which reason it's widely used in agriculture. Ho, staunchly opposed to genetic engineering for much of her career, considered the CaMV 35S promoter to be a "recipe for disaster."[2]

In Ho's view, CaMV 35S genes could stress the genetically modified plant by turning plant genes on or off far from the insertion site in the genome. The promoter might land next to a dormant viral gene and wake it up, generating a new superinfectious virus, or it might switch on a gene capable of producing a previously unknown deadly toxin. Similar concerns about GMO foods derived from plants containing CaMV 35S have been voiced by consumer groups.[3] Ho further speculated that if these foods were ingested by humans, the promoter could recombine with human viruses such as HIV, or jumping genes could activate oncogenes that cause cancer.

However, there's no scientific evidence to date that any of these fears about GMO foods have been realized, or are likely to be. Scientists from a prestigious British biotechnology laboratory, who included the discoverer of CaMV 35S, found there was no evidence to support any of Ho's arguments, and concluded that any risks involved in using the promoter are no

---

[1] The CaMV 35S promoter is part of the genome of the cauliflower mosaic virus (CaMV), a common plant virus that infects about 10% of cauliflowers and cabbages.
[2] Pringle, Peter. 2005. *Food, Inc.: Mendel to Monsanto — The Promises and Perils of the Biotech Harvest*. New York: Simon & Schuster, 96-97.
[3] Hansen, Michael K., Consumer Policy Institute/Consumers Union. 2000. "Genetic Engineering is Not an Extension of Conventional Plant Breeding; How Genetic Engineering Differs from Conventional Breeding, Hybridization, Wide Crosses and Horizontal Gene Transfer," January, https://consumersunion.org/wp-content/uploads/2013/02/Wide-Crosses.pdf. Accessed 12 July 2018.

greater than those encountered in conventional plant breeding.[1] They also pointed out that humans have been consuming the CaMV 35S promoter, in cauliflowers and cabbages infected with the CaMV virus, at levels more than 10,000 times greater than those in uninfected transgenic plants, throughout history. Other critics rejected the idea that the promoter could reactivate dormant human viruses or cause cancer, as a leap of logic and "pure fiction, and lies." One even suggested sarcastically, "Let's stop eating plants and animals altogether." [2]

Environmentalists have expressed a different fear about GMO crops: that jumping genes could escape from genetically engineered plants in the field into nearby weeds, endowing the weeds with inflated resistance to herbicides and pests. The resulting "superweeds" might become rampant and take over the environment. But so far no scientific evidence has emerged to substantiate such a damaging scenario. There's no indication that herbicide-resistant weeds in corn, soybean or cotton production regions have acquired that trait from genes that leapt from the crop.[3] In fact, it's entirely possible that the resistance results from natural pollination. And in a 10-year UK study ending in 2000, GMO crops such as rapeseed and sugar beet engineered for greater herbicide resistance showed no sign of becoming superweeds; the GMO plants didn't even survive for the length of the study.[4]

False Alarms About GMOs

Fear aside, many of us — in the U.S. and Canada at least — are probably unaware of just how much of what we eat every day is genetically modified food. In mid-2017, there were 11 GMOs approved by the USDA and FDA, including farmed salmon.[5] The approved crops include corn, which is the basic ingredient in many cereals, corn tortillas, corn starch and corn syrup, as well as feed for livestock and farmed fish; soybeans; canola; sugar beets; yellow squash and zucchini; bruise-free potatoes; and, most recently, nonbrowning apples. If you don't think you consume any of these or anything derived from them, be aware that ubiquitous french fries are most often cooked in GMO soybean or canola oils, the two vegetable oils also most

---

[1] Hull, R., S. N. Covey and P. Dale. 2000. "Genetically Modified Plants and the 35S Promoter: Assessing the Risks and Enhancing the Debate." *Microbial Ecology in Health and Disease*, 12, 1-5.

[2] Hodgson, John. 2000. "Scientists Avert New GMO Crisis." *Nature Biotechnology*, 18, 13.

[3] Hanson, Brad. 2014. "Can Herbicide Resistance Move from Crops to Weeds?" *UC Weed Science* blog, 28 July, http://ucanr.edu/blogs/blogcore/postdetail.cfm?postnum=14792. Accessed 12 July 2018.

[4] Crawley, M. J., S. L. Brown, R. S. Hails et al. 2001. "Transgenic Crops in Natural Habitats." *Nature*, 409, 682-683.

[5] 2016 "What Foods are Genetically Modified?" *Best Food Facts* blog, 15 August, https://www.bestfoodfacts.org/what-foods-are-gmo/. Accessed 12 July 2018.

frequently used by restaurants for other frying and in making pizza dough, as well as in numerous processed foods. And, while it's not a GMO as such, a small percentage of milk and other dairy products contain a genetically engineered hormone, recombinant bovine growth hormone (rBGH), injected into dairy cows to boost milk production. However, there's no scientific data showing that any of these GMO foods are deleterious to human health.

One of the first public outcries against GMO foods in the early 1990s involved a tomato modified with a fish gene. The tomato was intended to withstand extra cold temperatures, from either frost while growing or excessive refrigeration during transport, by engineering it with an antifreeze gene from the Arctic flounder. Yet the Frankenfish tomato never made it to market, not because it was shown to be unsafe for consumption but rather because it didn't exhibit the desired antifreeze effect. All the same, the very notion of tomatoes with fish eyes staring up at us from our plates generated an uproar at the time. The misinformed belief that supermarket tomatoes might be packed with fish genes was enough to scare activists and the public alike.

A few years later, development of another genetically modified food, in this case for animals, was also stopped in its tracks but for good reason, when plans to market a soybean containing a gene from Brazil nuts were shelved. Seed supplier Pioneer Hi-Bred International wanted to bolster the nutritional content of its soy-based animal feeds, which must normally be supplemented with an amino acid called methionine to promote adequate growth of the feeding animals. Because the Brazil-nut protein 2S albumin is very rich in methionine, Pioneer chose the 2S albumin gene to splice into the soybean genome. But mindful that Brazil nuts can cause strong allergic reactions in humans — though the specific allergen was unknown — and that soybeans intended for animals can't easily be separated from those destined for human consumption, the company commissioned testing of its transgenic soybeans for allergenicity. Sure enough, 2S albumin was found to be a human allergen and to be present in the transgenic soybeans, showing that genetic engineering can indeed transfer food allergens from one plant to another.[1] The positive test results, reported in 1996, would have required Pioneer to label its new product, under the FDA protocol for allergy testing in transgenic plants;[2] instead, the company dropped its marketing plans for the soybeans.

---

[1] Nordlee, Julie A., Steve L. Taylor, Jeffrey A. Townsend et al. 1996. "Identification of a Brazil-Nut Allergen in Transgenic Soybeans." *New England Journal of Medicine*, 334, 688-692.
[2] U.S. Department of Health and Human Services, Food and Drug Administration. 1992. "Statement of Policy — Foods Derived from New Plant Varieties," 29 May. *Federal Register*, 57, 22984-23004.

Marketing of yet another potential GMO food, potatoes genetically engineered to produce their own pesticide, was not pursued either following a major furor in the late 1990s. In this case, the idea was to make potatoes pest resistant by means of a gene borrowed from that harbinger of spring, the snowdrop flower. Despite its delicate appearance, the flower harbors a type of sugar-bearing protein known as a lectin that is toxic to certain kinds of preying insect. Biotechnology scientists saw the potential of the snowdrop gene inserted into a food plant for cutting back on pesticide spraying of the crop.

Hungarian biochemist Arpad Pusztai, a research scientist at Scotland's Rowett Institute, won a government grant to study the feeding of laboratory rats with lectin-modified transgenic potatoes. His ultimate goal was to ascertain whether the lectin gene in a food plant would be harmful to humans. Early experiments had shown that some lectins could penetrate animal cells and enter the bloodstream, just as they do in insects, although rats could safely eat the snowdrop lectin. In 1998, Pusztai, who was an international expert on plant lectins, staggered the world by announcing on British TV his preliminary finding that the transgenic potatoes stunted the rats' growth and degraded their immune systems. He went on to say in response to a question that he wouldn't eat GMO foods himself, and that it was "very, very unfair to use our fellow citizens as guinea pigs."[1] The Rowett Institute was flooded with phone calls from reporters, government officials and biotech companies across the globe and, in the aftermath, Pusztai was muzzled. In a defensive overreaction, the prominent Institute sealed Pusztai's laboratory, seized his notebooks and suspended him.

An internal inquiry concluded that Pusztai's data didn't support his conclusions, and that his experiments were too crude to justify any claims about snowdrop genes in potatoes. But a number of his colleagues rallied to his side and complained that Pusztai had been treated too harshly. He had, however, committed the then cardinal scientific sin of going public with unpublished data before having it peer reviewed, though that practice has become commonplace today. Nevertheless, a review by the UK's Royal Society also found that the lectin study was flawed in its design and analysis.[2] And to rub salt into the wound, the prestigious medical journal *The Lancet* weighed in by first allowing Pusztai to publish a watered-down version of

[1] ITV World in Action. 1998. "Eat Up Your Genes," Interview with Arpad Pusztai, 10 August. [See Enserink, Martin. 1998. "Institute Copes with Genetic Hot Potato." *Science*, 281, 1124-1125.]
[2] The Royal Society. 1999. "Review of Data on Possible Toxicity of GM Potatoes," 1 June, https://royalsociety.org/~/media/Royal_Society_Content/policy/publications/1999/10092.pdf. Accessed 12 July 2018.

his public statements,[1, 2] but then echoing the Institute and the Royal Society in publishing a Dutch paper, stating that Pusztai's results "do not allow the conclusion that the genetic modification of potatoes accounts for adverse effects in animals."[3]

While all these machinations in what has become known as the "poisoned rat debate" reflect the ongoing battle between anti-GMO campaigners and the establishment, a careful scientific analysis of Pusztai's data reveals that indeed his conclusions were mistaken. Although his experiments were criticized on several grounds including the use of too few animals, many critics and supporters alike overlooked a major deficiency in the study: the lack of adequate controls. The unmodified, nontransgenic potatoes fed to the control group of rats had been bred differently from the modified potatoes fed to the test group. Most likely unknown to Pusztai, who was neither a plant breeder nor a biologist, was that the process involved in inserting snowdrop genes into a potato plant can on its own cause marked changes in the genome, apart from any changes caused by the snowdrop gene itself. So his comparison between test and control rats was scientifically invalid unless they had both been fed potatoes having exactly the same breeding history, with or without the lectin modification — which they hadn't been.[4] Nevertheless, the controversy put a stop to further development of potatoes engineered with the snowdrop gene. The nonbruising potatoes mentioned above, which received FDA approval in 2015, are the only GMO potato currently marketed for human consumption.

In addition to unwarranted scares about GMO foods tested in the laboratory, there have been several false alarms over purported gene flow into the environment from commercial GMO crops. At the center of one alarm, which generated even more news coverage than Pusztai's experiments, was the monarch butterfly. The beautiful orange and black monarchs are wondrous and mysterious creatures, tens of millions of them migrating

---

[1] Horton, Richard. 1999. "Genetically Modified Foods: 'Absurd' Concern or Welcome Dialogue?" *The Lancet*, 354, 1314-1315.

[2] Ewen, Stanley W. B. and Arpad Pusztai. 1999. "Effect of Diets Containing Genetically Modified Potatoes Expressing *Galanthus Nivalis* Lectin on Rat Small Intestine." *The Lancet*, 354, 1353-1354.

[3] Kuiper, Harry A., Hub P. J. M. Noteborn and Ad A. C. M. Peijnenburg. 1999. "Adequacy of Methods for Testing the Safety of Genetically Modified Foods." *The Lancet*, 354, 1315-1316.

[4] Fedoroff, Nina V. 2011. "Pusztai's Potatoes — Is 'Genetic Modification' the Culprit?" *AgBioWorld* Ag-Biotech articles, http://www.agbioworld.org/biotech-info/articles/biotech-art/pusztai-potatoes.html. Accessed 12 July 2018.

In genetic engineering, there are several methods for culturing engineered *Agrobacterium* cells prior to regeneration of the genetically modified plant. A frequently used method involves the growth of calluses or blobs of tissue that generate shoots and roots. The variation seen in plants produced by this particular culturing process is called somaclonal variation.

thousands of miles each fall from the U.S. and southern Canada to winter habitats in Mexico and Florida. Monarch butterflies begin their lives on a milkweed plant, from eggs laid on the undersides of the upper leaves; the eggs hatch into tiny caterpillars or larvae that feed on the milkweed leaves — the only food that the larvae will eat — until they transform into a cocoon from which the adult butterfly emerges.

But milkweed often grows in and around cornfields. In 1999, Cornell University entomologist John Losey and two colleagues published a research paper claiming that pollen dispersed by corn plants genetically engineered with an insecticide gene from a soil bacterium known as Bt kills monarch butterfly larvae.[1] The paper immediately set the media world on fire, with multiple incendiary newspaper articles including one on the front page of the *New York Times*,[2] and another in which an entomologist called the monarch the "Bambi of the insect world."[3] The toxin-producing gene was at the time engineered into about 25% of the U.S. corn crop, and was used to modify other crops such as cotton as well; Bt insecticides are harmless to humans and popular with organic farmers, who regard them as natural. In corn, the Bt toxin is targeted at the European corn borer, a caterpillar that tunnels into the cornstalk causing the plant to topple.

Losey had carried out a simple experiment. After sprinkling damp milkweed leaves with Bt corn pollen until the coverage resembled what he'd seen on actual milkweed in a cornfield, he placed five three-day-old monarch caterpillars on each leaf and waited for them to feed. Understanding the importance of replication to the scientific method, he repeated the experiment four times. After four days of feeding, almost half the larvae exposed to Bt corn pollen were dead; those that survived had eaten less and were smaller than the control group fed on unpollinated milkweed leaves. The authors concluded that the results had "profound implications" for the conservation of monarch butterflies.

However, the study was quickly criticized, notably by another Cornell entomologist, Anthony Shelton. Comparing Losey and the worldwide reaction generated by his paper to Shakespearean characters who spread falsehoods and indulge in fantasy, Shelton and a coauthor cautioned that the public could be taken in by laboratory reports that "may not have any

[1] Losey, John E., Linda S. Rayor and Maureen E. Carter. 1999. "Transgenic Pollen Harms Monarch Larvae." *Nature*, 399, 214.
[2] Yoon, Carol Kaesuk. 1999. "Altered Corn May Imperil Butterfly, Researchers Say." *New York Times*, 20 May, http://www.nytimes.com/1999/05/20/us/altered-corn-may-imperil-butterfly-researchers-say.html. Accessed 12 July 2018.
[3] Weiss, Rick. 1999. "Gene-Altered Corn May Kill Monarchs." *Washington Post*, 20 May, http://www.washingtonpost.com/wp-srv/national/daily/may99/monarchs20.htm. Accessed 12 July 2018.

reality in the field or even in the laboratory." Commenting on the knee-jerk response of the public and policy makers to Losey's article, they went on to say they believed that "few entomologists or weed scientists familiar with butterflies or corn production give credence" to the work.[1] One of the scientific criticisms leveled at Losey by others was the slapdash method of controlling the uniformity of pollen sprinkled on the leaves, based only on his visual memory of field conditions. Furthermore, the laboratory conditions, such as they were, hardly simulated reality: the pollen density on a leaf in the field is unlikely to remain the same for long, since pollen blown onto the leaf by wind can be blown off again as well being washed away by rain. And at the time, there were several types of Bt corn available, each with different levels of Bt toxicity for monarch larvae.

Because of all the uncertainty and publicity over potential butterfly poisoning, the USDA and the biotech industry combined to launch a sweeping set of studies to examine all the factors likely to influence the impact of Bt corn pollen on monarch larvae. Tests were conducted both in the laboratory and in actual cornfields; exactly how toxic three varieties of Bt corn were to caterpillars was measured; the amount of corn pollen deposited on milkweed leaves in and away from cornfields was studied; the overlap in time between monarch egg laying and the 7–10 day period when corn plants shed their pollen was probed; and a host of other variables were investigated. The timing of pollen deposition was particularly important, as it wasn't known if the hatching of monarch larvae coincided with pollen shedding at all.

The studies found that the timing varies: the peak of monarch egg laying does indeed overlap with pollen shedding in Ontario and the northern U.S., but the two events don't coincide further south in Maryland. And only one of the three commercial Bt corn varieties was found to be toxic enough to kill monarch caterpillars, or even slow their growth, at the actual pollen densities in cornfields. That particular variety, known as event 176 hybrid, has since been withdrawn from the market. Unfortunately, the study results were published at the same time that the 9/11 terror attacks occurred, so got little immediate attention. Nevertheless, the research did come up with good news for monarch butterflies: it was estimated that only about 0.05% of the monarch population in Iowa, a major corn-growing state, is at risk from Bt

---

[1] Shelton, Anthony M. and Richard T. Roush. 1999. "False Reports and the Ears of Men." *Nature Biotechnology*, 17, 832. [See also Friedlander, Blaine. 1999. "Two Leading Researchers Take Issue with Three Recent Studies on the Effects of Genetically Engineered Crops." *Cornell Chronicle*, 9 September, http://www.news.cornell.edu/stories/1999/09/genetic-engineered-crop-studies-questioned. Accessed 12 July 2018.]

corn. That's one larva out of every 2,000 at risk of dying, and then only if the caterpillar eats enough pollen on the milkweed leaf.[1]

Another case of much ado about nothing was the discovery that native corn cobs grown in the Oaxaca mountains of southern Mexico were contaminated by alien genes found in genetically modified corn grown hundreds of miles away in the U.S. The discovery, reported in the journal *Nature* in 2001, was made by two visiting University of California, Berkeley researchers, ecologist Ignacio Chapela and his graduate student David Quist. Not only did Quist and Chapela claim that some of the local *criollo*[2] corn contained the CaMV 35S gene — the promoter frequently used in genetic engineering of plants that Ho had railed against — as well as a Bt insecticide gene, but also that the transferred genes were unstable. According to the authors, the transgenic DNA was found at random locations in the criollo genome and showed signs of having fragmented and moved around, rather like jumping genes. This conclusion raised the possibility that the intruder genes could be passed on to other types of plant. Control samples of criollo corn seeds from the time before GMO crops existed in showed no sign of either the CaMV 35S or Bt gene, they said.[3]

Pollen from American corn could not have flown over such a large distance and, at least officially, contaminated pollen could not have come from local sources as there had been a moratorium on growing genetically engineered corn in Mexico since 1998. However, Mexico has long imported corn from the U.S., much of which is GMO corn. Should Mexican subsistence farmers have chosen not to eat the imported corn but instead to plant it near criollo cornfields, in defiance of the ban, the criollo crop could indeed have been tainted by GMOs as the U.S. researchers maintained.

But it was Quist and Chapela's second claim about wandering genes that immediately generated controversy and alarm. If true, their finding would have challenged the very foundations of biotechnology and implied that genetic engineering was intrinsically imprecise and risky. For Mexico, where corn was first domesticated about 9,000 years ago, the finding would have imperiled their long history of maize cultivation and crop diversity.

Just like Pusztai's assertions about snowdrop-modified potatoes and Losey's paper on the poisoning of monarch butterflies, however, Quist and Chapela's claim about migrating genes was promptly denounced in scientific

---

[1] Fedoroff and Brown, 2004, 205-209.

[2] The *criollo* corn (maize) grown in southern Mexico is a variety cultivated by local farmers that is not crossbred with modern varieties, known as a landrace in the breeding community.

[3] Quist, David and Ignacio H. Chapela. 2001. "Transgenic DNA Introgressed into Traditional Maize Landraces in Oaxaca, Mexico." *Nature*, 414, 541-543.

circles. Leading the charge was *Nature* itself, announcing in 2002, after calling on Quist and Chapela to provide more data to back up their work, that their evidence was "not sufficient to justify" having published the paper. The journal's editorial stance was bolstered by criticism of the Berkeley research in two accompanying papers published in the same issue.[1] While not as damning as *The Lancet*'s retraction of Wakefield's 1998 paper on autism and the MMR vaccine, this was the first time in *Nature*'s 133-year history that the journal had withdrawn support for a published article. A major criticism made in the other two papers, as well as an editorial in another publication,[2] was that Quist and Chapela hadn't taken enough trouble to eliminate false positives in their results arising from sample contamination — for example, by the popular CaMV 35S promoter gene in Bt corn samples that they used as controls. Another scientific flaw was their failure to check for conventional hybridization of the criollo corn as an alternative to the wandering genes proposal.

Although Quist and Chapela's first claim about genetically contaminated corn was confirmed independently by the Mexican government, who blamed the interbreeding on illegal planting of imported corn, the researchers' conclusion about migrating transgenes continued to be vehemently attacked. The microbiologist author of one of the *Nature* critiques called the study a "testament to technical incompetence," while a member of the pro-GMO AgBioView online forum labeled it as "junk science" that should not have got beyond peer review.[3] Quist and Chapela subsequently acknowledged their technical problems and conceded that they were "backing off a bit,"[4] perhaps a tacit admission of their scientific mistakes. But the two authors were also unwittingly swept up into agricultural politics, in which emotions and activism often dominate the rational discourse of science. At least part of the acrimonious reaction to their Mexican research stemmed from their previous opposition to a $25 million deal in 1998 between their university and Swiss agribusiness giant Novartis, giving the company rights to the results

---

[1] Editorial note. 2002. *Nature*, 416, 600; Metz, Matthew and Johannes Fütterer. 2002. "Suspect Evidence of Transgenic Contamination." *Nature*, 416, 600-601; Kaplinsky, Nick, David Braun, Damon Lisch et al. 2002. "Maize Transgene Results in Mexico are Artefacts." *Nature*, 416, 601.

[2] Christou, Paul. 2002. "No Credible Scientific Evidence is Presented to Support Claims that Transgenic DNA was Introgressed into Traditional Maize Landraces in Oaxaca, Mexico." *Transgenic Research*, 11, iii—v.

[3] Avery, Alex. 2002. "Joint Statement from Scientists?" *AgBioWorld* newsletter, 21 February, http://www.agbioworld.org/newsletter_wm/index.php?caseid=archive&newsid=1361. Accessed 12 July 2018.

[4] Yoon, Carol Kaesuk. 2002. "Journal Raises Doubts on Biotech Study." *New York Times*, 5 April, http://www.nytimes.com/2002/04/05/us/journal-raises-doubts-on-biotech-study. html. Accessed 12 July 2018.

of plant genomics research supported by the grant. Quist and Chapela's objections pitted them against the GMO establishment and international food conglomerates.

The Anti-GMO Campaign

Ever since GMO foods first appeared on the market, an uneasy tension has existed between industrial agriculture and environmentalists. With its emphasis on large-scale farming, powerful new pesticides and herbicides, and genetic engineering of crop plants, agribusiness quickly became anathema to environmental activists and others who mourned the loss of small family farms and didn't want anyone meddling with their food. But the agricultural behemoths, along with many scientists, were arrogant and condescending, believing mistakenly that there was no need to explain the new biotechnology to a gullible public who would accept genetically modified foods without question. The resulting tussle between the two sides has obscured the underlying science and drawn out the development of potentially beneficial new crops.

Nothing epitomizes this clash more than golden rice, a rice genetically modified to contain beta-carotene, which is the naturally occurring pigment that turns carrots orange and corn yellow. Because beta-carotene produces vitamin A in the human body, golden rice was once seen as the answer to vitamin A deficiency in many parts of Asia and Africa, where rice is the staple food. Millions of poor children on both continents die or go blind each year from weakened immune systems caused by a lack of the vitamin. Beta-carotene is found in the leaves of the rice plant but not in the grain, unlike the yellow kernels in corn. Yet, while the idea of inserting the pigment into rice grains through genetic engineering was first conceived back in the 1980s, golden rice is still in development after more than 25 years of research.

The effort began with studies in rice biotechnology funded by the Rockefeller Foundation, aimed at identifying genes for beta-carotene production and splicing them into rice. Although it was thought originally that four transgenes are needed, it turns out that the job can be done with only two. The two genes were discovered as a result of collaboration between Swiss plant geneticist Ingo Potrykus and German biologist Peter Beyer, who met at a Rockefeller-sponsored seminar in 1992. Potrykus, who had performed groundbreaking research in genetic engineering of plants, is an altruist who devoted the later years of his career to helping developing countries grow more food. Beyer is an expert on beta-carotene in daffodils, which were the source of one of the required genes; the other came from a bacterium. But achieving success took years of hard work and it wasn't until

1999 that Potrykus and Beyer were able to produce grains of rice with the characteristic deep yellow hue of daffodils.

However, despite a wave of publicity about their accomplishment and a feature article in *Time* magazine with Potrykus on the cover,[1] what had already been a rising tide of protest around the world against GMO foods erupted into open hostility toward golden rice. Potrykus was accused of creating a Frankenfood and golden rice dismissed as "fool's gold" by Greenpeace, who claimed that a person would have to eat about 20 pounds of cooked golden rice per day to meet the daily requirement for vitamin A[2] — a claim since repudiated by the development, reported in 2005, of an improved golden rice with 20 times as much vitamin A-generating beta-carotene. Instead of GMOs, Greenpeace advocates lower-tech vitamin A supplementation as a more practical solution to the problem. Other detractors saw the genetic engineering feat simply as a Trojan horse, as a vehicle for launching other more profitable GMO crops in the developing world. Together with the Rockefeller Foundation, Potrykus and Beyer had targeted golden rice at the world's poor and underprivileged, with the intention of making the technology freely available to subsistence farmers.

While Potrykus was taken aback by the barrage of attacks on golden rice, he was still a long way from realizing his dream of overcoming vitamin A deficiency across the globe. The first grains of his miracle food were merely a "proof of concept" in inventors' terms, and many more steps remained before golden rice would be ready for growing and distribution to the world's undernourished masses. These included testing of the prototype rice to ensure it didn't contain any transferred daffodil allergens, like the Brazil nut allergen found in Pioneer's transgenic soybeans; removal of special "marker" genes that had been necessary in the laboratory experiments, but were not required in field crops; and breeding of the new genes into strains of rice that could be easily grown in the tropics and subtropics, where the need for golden rice was greatest.

But if Potrykus was going to give the technology away to the poor, he would need to patent his invention. It came as a shock to the two researchers that their work had potentially infringed on no fewer than 70 patents belonging to 32 different companies and universities, all of which

---

[1] Nash, J. Madeleine. 2000. "This Rice Could Save a Million Kids a Year." *Time*, 31 July, 38-46, http://content.time.com/time/magazine/article/0,9171,997586-1,00.html. Accessed 12 July 2018.

[2] Greenpeace. 2001. "Genetically Engineered 'Golden Rice' is Fool's Gold." *AgBioWorld* newsletter, 9 February, http://www.agbioworld.org/newsletter_wm/index.php?caseid=archive&newsid=950. Accessed 12 July 2018.

would have to be licensed before Potrykus and Beyer could proceed with commercialization of their product. Potrykus lamented that:

> It seemed to me unacceptable, even immoral, that an achievement based on research in a public institution and exclusively with public funding and designed for a humanitarian purpose was in the hands of those who had patented enabling technology earlier.... It turned out that whatever public research one was doing, it was all in the hands of industry (and some universities).[1]

Realizing, however, that licensing was their only choice but they would need the assistance of a commercial partner, Potrykus and Beyer cut a deal with British pharmaceutical company AstraZeneca, which had funded a portion of the research and had rights to their invention. Under the deal, AstraZeneca would allow the two scientists to give golden rice seeds away to farmers in developing countries who made less than $10,000 per year, in exchange for exclusive marketing rights to golden rice in the industrialized world. Once the deal was announced, other agribusiness corporations such as Monsanto joined in by giving Potrykus and Beyer free licenses to their patents as well.

Nevertheless, golden rice is still years away from production. The crossbreeding necessary to insert the magic daffodil and bacterium genes into suitable varieties of rice is being conducted at the nonprofit International Rice Research Institute in the Philippines, the very place where rice breeders first came up with the idea of golden rice. According to a recent study from Washington University in St. Louis, researchers at the Institute have so far been unable to develop golden rice strains with yields as high as those of non-GMO strains already being grown by farmers. And it's still unknown if the beta-carotene in golden rice can even be converted to Vitamin A in the bodies of badly malnourished children.[2]

Apart from the scientific obstacles to the development of a viable golden rice, progress has also been slowed by environmental politics. During the period that Potrykus and Beyer were working slavishly on golden rice transgenes in the 1990s, protests against GMO foods were taking place all over Europe. In France and Ireland, protesters vandalized seed stores and ripped up test plots. Austria, Luxembourg and Norway banned the planting of genetically engineered corn — a prelude to the current ban on growing GMO crops in more than half the countries in the European Union — and

---

[1] Potrykus, Ingo. 2001. "Golden Rice and Beyond." *Plant Physiology*, 125, 1157-1161.
[2] Everding, Gerry. 2016. "Genetically Modified Golden Rice Falls Short on Lifesaving Promises." *The Source* blog, Washington University in St. Louis, 2 June, https://source.wustl.edu/2016/06/genetically-modified-golden-rice-falls-short-lifesaving-promises/. Accessed 12 July 2018.

food stores in Britain stopped selling GMO products at the time. In the U.S., where the FDA had in 1996 given its blessing to the marketing of GMO crops, a 1998 lawsuit accused the agency of negligent oversight of genetically modified foods.

The ferocity of the protests in Europe prompted the Swiss government to construct a bombproof greenhouse for Potrykus to continue his work on golden rice and other transgenic crops, without fears of disruption.[1] Ironically, it is mostly fears about food safety that motivate anti-GMO activists, so there is fear on both sides of the GMO debate. The intensity of the European opposition to genetically modified foods had been exacerbated by the scare over mad-cow disease in the early 1990s. Later in the decade, consumer anger over both the mad-cow episode and GMOs was taken out on the agribusiness giants, especially Monsanto. Unwisely, Monsanto chose to start marketing its new GMO crops in Europe at just that time and had already antagonized North American customers with its aggressive business practices.

The patent on Monsanto's popular herbicide Roundup, a best-seller for more than 20 years, was approaching expiration when the company introduced Roundup Ready seeds in 1996. The seeds were genetically engineered with a new gene that made crop plants resistant to the herbicide, a revolutionary advance that meant farmers could now use Roundup to kill weeds while the crop was growing, instead of only before planting. Current Roundup Ready crops include canola, corn, soybean, sugar beet and wheat.

Glyphosate, the active ingredient in Roundup, was recently in the crosshairs of anti-GMO activists when the European Commission voted in 2017 to limit the chemical's reauthorization in Europe to only five years. Widely used around the world for more than 40 years to combat weeds and to develop sustainable farming techniques such as no-till, glyphosate has also drawn the attention of regulators as possibly carcinogenic. But, while multiple European regulatory bodies such as the European Chemicals Agency have concluded that the chemical is safe for humans and animals, the International Agency for Research on Cancer (IARC) in 2015 classified glyphosate as a potential carcinogen. This conclusion underpinned the Commission's decision to renew glyphosate approval for less than the customary 10 or 15 years; many individual European countries have said they'll terminate its use within three years. In the U.S., however, the EPA has disputed the IARC finding and in 2018 declared glyphosate "not likely to

---

[1] Rao, C. Kameswara. 2007. "Bombs, Bunkers and Golden Rice." *Foundation for Biotechnology Awareness and Education (FBAE)* Special Topics: Views, 27 December, http://www.fbae. org/2009/FBAE/website/special-topics_views_bombs.html. Accessed 12 July 2018.

be carcinogenic to humans."[1] Nevertheless, Monsanto faces approximately 4,000 lawsuits across the country filed by victims of non-Hodgkin's lymphoma, a cancer the victims claim was caused by spraying the company's Roundup herbicide. In a preliminary court hearing in March 2018, dubbed "Science Week," both Monsanto and the plaintiffs, who include residential users as well as farmers, maintained that the scientific evidence was on their side.[2]

Monsanto has also come under fire over its Roundup Ready seeds. In 1998, the agricultural colossus launched an unsuccessful bid to buy Delta and Pine Land Company, the leading U.S. cotton seed supplier that had recently patented a genetic engineering technology to make plant seeds produced by a crop sterile. If adopted by Monsanto, the new technology would have forced its customers to purchase new seeds each year and prevented them planting seeds from the previous year's crop, as they had been able to do in the past — despite this practice becoming illegal for patented seeds after 1985. But, while Monsanto in 2007 finally acquired Delta and Pine Land along with the sterile seed patent, the company has undertaken not to commercialize "terminator" seeds.[3]

Nonetheless, while Roundup Ready seeds guaranteed Monsanto continued profits, chairman Robert Shapiro was determined to complete the transition he had already embarked on, in taking Monsanto from its tarnished environmental past in the chemicals industry to what he saw as a more humanitarian role in biotechnology. After acquiring a smaller biotechnology company, Shapiro bought three seed companies which, together with subsequent acquisitions after 2000, have made Monsanto (now Bayer) the world's largest seed conglomerate. By 1999, Monsanto's stock price had doubled and the company seemed poised to become the dominant force in GMO crops and foods. However, the GMO landscape was rapidly changing. Because of the backlash against GMOs in Europe, Monsanto's hopes of marketing its enormous seed inventory there had been dashed, and two of its largest competitors — Novartis and AstraZeneca — were about to downsize and merge their two separate agribusiness divisions into a single company. In Japan, the government had received petitions opposing genetically modified

---

[1] U.S. Environmental Protection Agency. 2018. "Draft Human Health and Ecological Risk Assessments for Glyphosate," https://www.epa.gov/ingredients-used-pesticide-products/draft-human-health-and-ecological-risk-assessments-glyphosate. Accessed 12 July 2018.

[2] U.S. Right to Know. 2018. "The Monsanto Papers: Roundup (Glyphosate) Cancer Case Key Documents & Analysis," https://usrtk.org/pesticides/mdl-monsanto-glyphosate-cancer-case-key-documents-analysis/. Accessed 12 July 2018.

[3] Monsanto. 2017. "Myth: Monsanto Sells Terminator Seeds," 10 April, http://www.monsanto.com/newsviews/pages/terminator-seeds.aspx. Accessed 12 July 2018.

crops and demanding that GMO foods be labeled. Baby food manufacturers Gerber and Heinz had given in to Greenpeace pressure and stopped using genetically modified soy and corn in their products.[1]

With the FDA about to begin hearings on the use and safety of genetically altered foods, a contrite Shapiro appeared on a video screen at the 1999 annual fall meeting of Greenpeace, and apologized for Monsanto's corporate arrogance and insensitivity to the protests over GMOs, "... because we thought it was our job to persuade. But too often we forgot to listen." Greenpeace director Lord Melchett, accusing Shapiro of being a bully, claimed that the vast majority of the anti-GMO movement were not opposed to science but simply suspicious of big science and agribusiness.[2] Indeed, Monsanto's heavy-handed tactics and its monopoly of the Roundup Ready seed market, essentially forcing farmers to purchase Monsanto seeds if they want to use Roundup herbicide, are prime examples of how the agribusiness giants have misused GMO science for commercial gain.

Melchett's declaration was deceptive, however, as many anti-GMO activists have little time for science. A prominent example is the UK's Prince Charles, who believes that "this kind of genetic modification takes mankind into realms that belong to God, and to God alone."[3] Charles, a small-scale organic farmer himself, has also said that agribusiness companies have been "conducting an experiment with nature ... which has gone 'seriously wrong' ." Although he advocates more cooperative farms in the British agricultural system, Charles denies that he's trying to go back to the past, maintaining that scientists are putting too much pressure on nature with genetic engineering that is threatening the diversity of staple crops such as soy, corn, wheat and rice.[4]

But this is nonsense according to botanist and environmentalist Peter Raven, former director of the Missouri Botanical Garden. In Raven's view, selected transgenes that enhance resistance to pests and disease are likely to promote biodiversity by increasing productivity and improving habitat quality. The most serious threats he sees to both the diversity of Mexican

[1] Specter, Michael. 2000. "The Pharmageddon Riddle." *New Yorker*, 10 April, 58-71, http://www.michaelspecter.com/2000/04/the-pharmageddon-riddle/. Accessed 12 July 2018.

[2] Vidal, John. 1999. "We Forgot to Listen, Says Monsanto." *The Guardian*, 6 October, https://www.theguardian.com/science/1999/oct/07/gm.food. Accessed 12 July 2018.

[3] Charles, Prince of Wales. 1998. "The Seeds of Disaster." *The Telegraph*, 8 June, https://www.princeofwales.gov.uk/speech/article-prince-wales-titled-seeds-disaster-daily-telegraph. Accessed 12 July 2018.

[4] Randall, Jeff. 2008. "Prince Charles Warns GM Crops Risk Causing the Biggest-Ever Environmental Disaster." *The Telegraph*, 12 August, http://www.telegraph.co.uk/news/earth/earthnews/3349308/Prince-Charles-warns-GM-crops-risk-causing-the-biggest-ever-environmental-disaster.html. Accessed 12 July 2018.

maize and biodiversity in general are habitat destruction, which includes a large portion of the world's land devoted to agriculture and grazing; urbanization; invasive weeds and pests; and the exploitation of certain animals and plants, for example by overfishing the oceans or by harvesting exotic plants for medicine. Raven points out that gene flow has been a feature of the entire history of crop evolution, by both artificial selection and hybridization, long before genetic engineering entered the picture.[1]

There are currently 28 nations around the world that allow the cultivation of genetically modified crops — 20 developing countries and 8 industrialized countries including the U.S., Canada, Brazil and Argentina.[2] In the U.S., GMO crops accounted for 92% of the total corn acreage in 2016.[3] Worldwide, however, by far the largest GMO crop is soybeans, with corn the next largest followed by cotton. But the growing of genetically engineered crops is banned in more than 30 other nations, including Russia and over half the membership of the European Union, even though most of these countries allow the import of GMO foods, primarily for animal feed. In the UK, England has given the green light to research on genetically modified crop varieties for many years, and has approved field trials of GMO wheat and potatoes, though no commercial GMO crops have been grown to date. Scotland, Wales and Northern Ireland have all joined the countries banning GMO crops altogether.

Rejection Of Science

What is striking about the GMO food story is the faithful adherence to the scientific method by those involved in the development of genetic engineering, but the much less resolute following of scientific principles on occasion by GMO critics. Mendel's pioneering experiments in genetics, McClintock's diligent research on jumping genes, the creation of golden rice by Potrykus and Beyer, and countless laboratory and field studies of the impact of various transgenes all exemplify scientists embracing the central features of the scientific method: observation, hypothesis and reason. As

---

[1] Raven, Peter H. 2014. "GM Crops, the Environment and Sustainable Food Production." *Transgenic Research*, 23, 915-921, http://people.forestry.oregonstate.edu/steve-strauss/sites/people.forestry.oregonstate.edu.steve-strauss/files/Raven%20-%20Transgenics%20and.Env%20-%20Transg%20Res%202013.pdf. Accessed 12 July 2018.

[2] Genetic Literacy Project. 2016. "GMO FAQ: Where are GMOs Grown and Banned?," https://gmo.geneticliteracyproject.org/FAQ/where-are-gmos-grown-and-banned/. Accessed 12 July 2018.

[3] U.S. Department of Agriculture, Economic Research Service. 2017. "Adoption of Genetically Engineered Crops in the U.S.: Recent Trends in GE Adoption," https://www.ers.usda.gov/data-products/adoption-of-genetically-engineered-crops-in-the-us/recent-trends-in-ge-adoption.aspx. Accessed 12 July 2018.

we've seen, Frankenfish tomatoes and soybeans enhanced with a Brazil-nut protein were both terminated in the laboratory, based on rational assessments of the scientific evidence.

Such high regard for the scientific method contrasts with the bias and sloppiness of other scientists that led to the false alarms about GMO foods and crops discussed earlier in the chapter. Pusztai allowed his personal convictions that GMOs are harmful to bias his test results on potatoes spliced with snowdrop genes, by failing to ensure that the potatoes fed to both test and control rats had been bred the same way. Losey likewise jumped to a premature conclusion on the impact of Bt corn pollen on the monarch butterfly, based on a simplistic and carelessly conducted experiment that didn't simulate real-world conditions. It took a comprehensive, more carefully thought-out suite of studies to demonstrate that monarchs are endangered far less by genetically engineered corn than by other threats such as the mowing of roadsides or the spraying of herbicides. And Quist and Chapela's claim that genes had wandered from genetically modified U.S. corn grown in Mexico to neighboring native criollo corn was flawed for several reasons, not the least of which was a lack of attention to elimination of false positives in their test data. Above all, what was missing in each of these three cases was the questioning inherent in scientific skepticism, a crucial element of the scientific method. Had these scientists been more skeptical, their research might have produced quite different results.

However, science is under attack far more from its complete rejection by anti-GMO nonscientists, members of the public at large, than it is from being abused by the shortcomings of research scientists carrying out experiments on GMOs. This rejection arises primarily from fear — fear of the *unknown*, unlike the fear spread by anti-vaccinationists that we examined in Chapter 6, based on the *known* side effects of vaccination. The Frankenfood images conjured up by anti-GMO activists substitute fear for reason and give credence to the assertion that GMO foods are unnatural.

The rejection of science is all too apparent in the statements of politicians and nonscientific celebrities who have taken up the anti-GMO cause. Scotland's 2015 decision to ban the cultivation of GMO crops wasn't based on scientific evidence, admitted First Minister Nicola Sturgeon.[1] And Prince Charles not only invokes God but has also stated that, like Pusztai — who to this day has a large following among anti-GMO activists — he has no wish to eat genetically modified food nor "knowingly offer this sort of produce to my

---

[1] Gardham, Magnus. 2015. "GM Crops Ban Not Based on Scientific Considerations, Admits Sturgeon." *Herald Scotland*, 10 September, http://www.heraldscotland.com/news/13713650.GM_crops_ban_not_based_on_scientific_considerations__admits_Sturgeon/. Accessed 12 July 2018.

family or guests."[1] U.S. Congressman Dennis Kucinich, a long-time advocate of labeling GMO foods, insists that genetic engineering is a violent process having "nothing to do with the ways of nature," and that we're all part of "a grand experiment now in our food."[2] Potrykus, in a commentary on the challenges facing the development of golden rice, including the anti-science stance of politically motivated protesters, complained that rational arguments were "poor ammunition against the emotional appeals of the opposition."[3]

The turn against GMO science is probably most prominent in Europe where, as discussed earlier, over half the 28 countries (as of 2017) in the European Union have banned the cultivation of GMO crops. The European Union's position on GMOs stems from the Precautionary Principle which, as we saw in Chapter 5, advocates caution or even inaction in the face of scientific uncertainty.[4] However, the European stance is not what it seems, since large quantities of GMO animal feed are imported into Europe for the livestock industry. Ironically, it was Belgian molecular biologists Marc Van Montagu and "Jeff" Schell (1935–2003) who pioneered the *Agrobacterium* bacterial truck method for transferring foreign genes into plants. But today the lead in genetic engineering has passed to the U.S., and European farmers have gone back to traditional chemical spraying, which includes Roundup as we've seen, to combat pests and disease. Growing GMO crops is also prohibited in sub-Saharan Africa, denying farmers there the benefits of drought-tolerant and disease-resistant GMO varieties.[5]

The principal fear of those opposed to GMOs is the safety of genetically engineered foods, a concern magnified to some extent by the cavalier attitude of the agribusiness companies in the 1990s, when GMO foods were introduced to the marketplace. While there is no scientific evidence to date that GMOs are deadly or even unhealthy for humans, this doesn't mean that harmful effects can be categorically ruled out, as noted by the Royal Society in its review of Pusztai's data on lectin-modified potatoes. Absence of evidence doesn't *prove* that GMO foods are safe beyond all possible

[1] Charles, Prince of Wales. 1998. "The Seeds of Disaster." *The Telegraph*, 8 June, https://www.princeofwales.gov.uk/speech/article-prince-wales-titled-seeds-disaster-daily-telegraph. Accessed 12 July 2018.

[2] Media Roots. 2013. "Dennis Kucinich Transcript: Iraq, Accountability & GMOs," 25 March, http://mediaroots.org/dennis-kucinich-on-bts-iraq-truth-accountabilitydrones-gmos/. Accessed 12 July 2018.

[3] Potrykus, Ingo. 2001. "Golden Rice and Beyond." *Plant Physiology*, 125, 1157-1161.

[4] Science and Environmental Health Network. 2018. "Precautionary Principle, Missoula Statement: Conservation Decisions in the Face of Uncertainty," November 2000, http://www.sehn.org/amsci.html. Accessed 12 July 2018.

[5] Lynas, Mark. 2015. "With G.M.O. Policies, Europe Turns Against Science." *New York Times*, 24 October, https://www.nytimes.com/2015/10/25/opinion/sunday/with-gmo-policies-europe-turns-against-science.html?_r=1. Accessed 12 July 2018.

doubt. Nevertheless, regulatory authorities have come up with practical definitions of food safety based on experience, such as the FDA's phrase "generally recognized as safe (GRAS)," which requires of any food additive "a substantial history of consumption for food use by a significant number of consumers" before the additive can be granted GRAS status.[1]

By the FDA criterion, GMO crops and foods are safe to eat, concluded a 2016 study by the National Academies of Sciences, Engineering, and Medicine. The study authors found no substantial evidence that the risk to human health was any different for current commercial GMO crops than for their traditionally crossbred counterparts, nor was there any scientific backing for the assertion that GMO plantings cause environmental devastation. The report drew on three types of testing to evaluate genetically engineered crops and the foods derived from them: animal-feeding studies, compositional analysis of GMO plants and allergenicity testing. The animal-feeding tests provided "reasonable evidence" that eating GMO foods didn't harm animals, and the differences in the nutrient and chemical compositions of GMO plants compared to similar non-GMO varieties fell within the range of natural variation for non-GMO crops. As for specific health problems such as allergies or cancer possibly caused by eating genetically modified foods, the study authors found no difference between the results of epidemiological studies conducted in the U.S. and Canada, where GMO foods have been consumed since the late 1990s, and similar studies in the UK and Europe, where very few GMO foods are consumed.[2]

Nevertheless, the anti-GMO movement continues to insist that eating genetically altered foods is fraught with danger. Steven Druker, a public interest attorney who spearheaded the 1998 lawsuit on GMOs against the FDA, has recently written a book not only accusing the FDA and others of scientific fraud in covering up the alleged hazards of GMO crops, but also claiming that the media, the general public and even American presidents have been duped by the same elaborate fraud.[3] However, the book draws heavily on Pusztai's flawed research on transgenic potatoes, rehashes debunked conspiracy theories and doesn't present any scientific evidence in

---

[1] U.S. Department of Health and Human Services, Food and Drug Administration. 2018. "Ingredients, Packaging & Labeling: Generally Recognized as Safe (GRAS)," https://www.fda.gov/food/ingredientspackaginglabeling/gras/. Accessed 12 July 2018.

[2] U.S. National Academies of Sciences, Engineering, and Medicine. 2016. *Genetically Engineered Crops: Experiences and Prospects.* Washington, DC: National Academies Press, 16-19, https://www.nap.edu/download/23395. Accessed 12 July 2018.

[3] Druker, Steven M. 2015. *Altered Genes, Twisted Truth: How the Venture to Genetically Engineer Our Food Has Subverted Science, Corrupted Government, and Systematically Deceived the Public.* Salt Lake City: Clear River Press.

support of the inherent risks of GMO crops.[1] Most scientists, therefore, have not taken the book seriously.

In addition to its safety findings, the NAS report observed that advances in both conventional hybridization and genetic engineering are blurring the once clear distinction between these two processes for growing better crops. One recent advance, which is a combination of crossbreeding with genetic modification, is a gene-editing tool that enables existing genes in a plant to be modified or deleted[2] — instead of splicing alien genes into the genome, as had been done previously. Still, both crossbreeding and genetic engineering can result in unintended genetic effects as I discussed before, so both types of crop raise safety issues in foods. Because the two breeding processes are converging, the report authors called for regulation of the final *product* rather than the *process*. The current practice in the U.S. is process regulation, GMO crops being the province of the EPA and USDA, while food safety is overseen by the FDA. The USDA says, however, that the new gene-edited food crops don't need to be regulated any differently than unaltered crops.[3] The NAS report also recommended a tiered approach to safety testing, in which DNA analysis technologies are used to evaluate the risks to human health or the environment of a new plant produced by either process, and safety testing is then conducted only on those plant varieties that reveal potential hazards.

A less positive finding of the report was that crop yields of corn, soybeans and cotton have not increased any faster since the introduction of GMO varieties than they did historically in the U.S., even though the holy grail of genetic engineering is to boost yields. Commented Michael Hansen, senior scientist at Consumers Union, "Despite industry claims, these crops are clearly not the answer to world hunger."[4] A similar lack of enhanced yields has beset the golden rice program, as we've already noted.

---

[1] Chassy, Bruce M. 2015. "Steven Druker: Twisted Truth in Altered Genes Book." *GMO Answers* blog, 20 July, https://gmoanswers.com/studies/steven-druker-altered-genes-twisted-truth-book. Accessed 12 July 2018.

[2] The tool, known as Clustered Regularly Interspaced Palindromic Repeats (CRISPR), enables changes to be made easily to the genome of any organism, with greater precision than ever before. Initially utilized in medicine to help fight disease, CRISPR has more recently been applied to agriculture, resulting in products such as nonbrowning mushrooms and fungus-resistant wheat.

[3] U.S. Department of Agriculture, Animal and Plant Health Inspection Service (APHIS). 2016. Letter from Michael J. Firko, APHIS Deputy Administrator, to Yinong Yang, Pennsylvania State University, 13 April, https://www.aphis.usda.gov/biotechnology/downloads/reg_loi/15-321-01_air_response_signed.pdf. Accessed 12 July 2018.

[4] Pollack, Andrew. 2016. "Genetically Engineered Crops are Safe, Analysis Finds." *New York Times*, 17 May, https://www.nytimes.com/2016/05/18/business/genetically-engineered-crops-are-safe-analysis-finds.html?em_pos=large&emc=edit_nn_20160518&nl=morning-briefing&nlid=73839238&_r=1. Accessed 12 July 2018.

Once again, politics are inextricably intertwined with science in the GMO food debate, just as they are in the nutritional science debate over dietary fat and in climate science over the contentious issue of global warming. Two French scientists in fact admonished anti-GMO activist Ho for mixing her science with politics. In debunking her claim that the widely used CaMV 35S promoter could behave like a jumping gene and move around the host plant's genome, possibly reactivating sleeping viruses or causing cancer, they wrote, "Considering the complexity of the debate concerning GMOs,... it is essential to take care to make the distinction between scientific questions raised for reasons that are primarily political, and the truly scientific ones...."[1] In a previous section, we saw how Quist and Chapela became embroiled in agricultural politics when they claimed to have actually found wandering transgenes in Mexican corn. By venturing too far into the political realm, both Ho, and Quist and Chapela, did science a disservice.

But for all the abuses of science by GMO scientists, and the complete rejection of science by 40% or more of the populace, genetic engineering is remarkably free of the *ad hominem* attacks that have characterized the continental drift controversy in geology (Chapter 2), the global warming debate in climate science at times (Chapter 5), and the fight over mandatory vaccination in medicine (Chapter 6). This difference may simply reflect the fact that most scientists in genetic engineering, both pro-GMO and anti-GMO, are "insiders" in the sense that they have scientific credentials in closely related fields — that is, they belong to the same club. Wegener, the father of continental drift theory, but with a background in meteorology rather than geology, threatened the early-20th-century geology establishment who saw him as an outsider, and so attacked him venomously. Likewise, late 20th-century climatologists impugned skeptics who disputed that climate change is largely due to human activity and were often not climate scientists. In the case of GMOs, the rage of anti-GMO activists has been directed mostly at agricultural conglomerates and GMO crops growing in the fields, rather than individuals. Even the critics of Ho and other activist scientists have stopped short of vicious personal attacks.

At the consumer level, fear of GMOs manifests itself in ongoing debate over labeling of GMO foods. The 2016 NAS report, which designated GMO crops and foods as safe, said there was no public health reason to label foods

[1] Morel, Jean-Benoît and Mark Tepfer. 2000. "Pour une Évaluation Scientifique des Risques: le Cas du Promoteur 35S (Are There Potential Risks Associated with Use of the Cauliflower Mosaic Virus 35S Promoter in Transgenic Plants?)." *Biofutur* 201, 32-35, English translation, http://citeseerx.ist.psu.edu/viewdoc/download?doi=10.1.1.569.3214&rep=rep1&type=pdf. Accessed 12 July 2018.

containing GMOs, though labeling could be justified for other reasons such as consumers' right to know. But if labeling is only voluntary, consumers would not know whether a particular product contained GMO ingredients or not, so could not make an informed choice. Mandatory labeling of GMO foods, on the other hand, would overcome this objection but could be interpreted as a warning, suggesting that GMO ingredients aren't safe when in fact there isn't any scientific evidence to support such a claim.

The U.S. Congress passed legislation in 2016 mandating the labeling of foods containing GMO ingredients; the USDA was finalizing the regulations in 2018. This followed a significant uptick in voluntary labeling of non-GMO foods over the previous 10 years. The new legislation brings the U.S. in line with many other countries, although the mandatory label is not required to display the GMO content in words: alternatives include a picture or a QR code that can be scanned with a smartphone.[1] But at least a GMO label is more scientific than calling the product a Frankenfood.

---

[1] U.S. Department of Agriculture, Agricultural Marketing Service. 2018. "National Bioengineered Food Disclosure Standard," 4 May. *Federal Register*, 83, 19860-19889.

Chapter 8. The Assault on Science

It should be clear from the previous chapters that science today is indeed under assault and that the voices of unreason are becoming louder. At the very least, the fabric of science is fraying badly if not actually torn. We've looked at several hot-button issues where science has been abused, sidelined or ignored, but these are far from being the only examples. In this final chapter, we'll identify the forces arrayed against the modern scientific enterprise, in order to understand just why science is currently ailing.

Mistreatment of Evidence

One of the most obvious signals that something is wrong with modern science is lack of regard for the importance of empirical evidence. All too often, logic is thrown out the window and scientific evidence is misinterpreted, misrepresented or simply cast aside. Yet the collection of evidence by observation is the very basis of the scientific method, as I've emphasized throughout the book. Mistreatment of evidence is a major abuse of science.

In Chapter 2, we saw how acceptance of Wegener's theory of continental drift was delayed for many decades. Initially, this was because geologists of the day were unwilling to take seriously his circumstantial and somewhat sparse evidence of matchups in coastlines, animal and plant fossils, mountain chains and glacial deposits, preferring to hang on to the consensus theory of a contracting earth to explain these disparate phenomena. Later, after Wegener's unfortunate early death, the geological community still refused to pay much attention to the more plentiful direct evidence for continental drift provided by rock magnetism. Eventually, the observational evidence of magnetic striping on the seafloor and the emergence of plate tectonics theory brought the geological world to its senses and Wegener's theory was accepted.

Chapter 3 revealed that during the late 19th century, Darwin too had to battle for acceptance of his theory of evolution. His proposed mechanism of natural selection didn't become part of the scientific mainstream until well into the 20th century, after the new science of genetics and its offshoots gained respectability. In Darwin's lifetime, all his empirical evidence for evolution was largely neglected, mainly because it conflicted with religious thinking and triggered a resurgence of opposing creationist beliefs. But despite widespread adherence to such beliefs even today, especially in the U.S., there is no scientific evidence whatsoever to support creationism. Young-Earth creationism twists the evidence of the fossil record to conform to the beliefs of its faith, derived from a literal interpretation of the Bible, in order to come up with the pseudoscience of flood geology. Intelligent design, the other principal form of modern creationism, is also faith-based but ignores scientific evidence completely. Its concepts of irreducible complexity and complex specified information both exemplify the philosophical argument from ignorance: highly complex biological systems must have been created by an intelligent designer, because we can't fathom how nature could have produced them on its own.

In the case of Keys' dietary fat hypothesis discussed in Chapter 4, the evidence supporting the hypothesis is weak. Keys' conviction that an excess of saturated fat in the diet is the main cause of death from coronary heart disease is no longer widely accepted. Although initial evidence from the Seven Countries Study and the earlier Framingham Heart Study suggested that Keys was right, subsequent follow-ups of both studies (after 25 and 30 years, respectively), together with results from the Women's Health Initiative, turned that conclusion on its head. The earlier evidence had been misinterpreted, a finding endorsed more recently by meta-analyses and systematic reviews that have pooled the evidence for and against the diet–heart hypothesis from a multitude of studies. In the public arena, the USDA simply ignored the contrary evidence altogether, promoting a low-fat diet in its 1980 *Dietary Guidelines*.

Similarly, the actual empirical evidence for a substantial human contribution to global warming is flimsy and frequently misrepresented. As explained at length in Chapter 5, the whole case for catastrophic consequences of man-made climate change — the case which the 195 parties to the toothless Paris Agreement, along with many of the world's scientific societies and national academies, have signed on to — is based entirely on artificial computer models, not observational evidence. The only way that climate change can be linked to human activity is through the models; the observations alone don't prove it comes from humans. However, not only is computer model output *not* observational evidence, but the predictive skills of current models are poor. The models failed to predict the global warming pause, wrongly predict a hot spot in the upper atmosphere that isn't there, haven't captured the present hiatus in previously rising ocean heat, and don't accurately reproduce sea level rise. Are the models evidence that our CO_2 emissions have a major effect on climate? Hardly.

An abundance of observational evidence does exist for the effectiveness and safety of vaccination (Chapter 6). Yet a small but vocal minority of anti-vaccinationists turn a blind eye to the evidence and claim erroneously that vaccines have a high rate of disabling side effects. And, though there is no scientific evidence that GMOs are any more harmful to human health than conventionally crossbred plants (Chapter 7), a surprisingly high 40% of the U.S. public, and perhaps a higher percentage in Europe, believes otherwise. As American evolutionary biologist Jerry Coyne has stated, "The fear of GMOs is like creationism: an unfounded belief based not on facts, but on a form of faith: genetically unmodified food is better."[1] A similar creationist-like belief lies behind fear of vaccines: the disease is better than the vaccine.

The disregard for the importance of empirical evidence reflects the erosion of public trust in science — a subject discussed extensively over the last 50 years and the topic of a 2015 workshop organized by the National Academies of Sciences, Engineering, and Medicine.[2] A large percentage of the public at large, used to being lied to by politicians, governments and even world leaders, simply no longer trusts scientific evidence.

Corrosion And Corruption Of Science

A closely related issue and another signal that science is under attack is the growing recognition of a crisis in reproducibility of scientific evidence. As discussed in Chapter 1, replication of scientific data, preferably by independent investigators in different laboratories, is not only an essential part of the scientific method but is also crucial to the credibility of science itself.

---

[1] Coyne, Jerry. 2014. "The Science Guy Goes After GMOs." *Why Evolution is True* blog, 8 November, https://whyevolutionistrue.wordpress.com/2014/11/08/the-science-guy-goes-after-gmos/. Accessed 12 July 2018.

[2] U.S. National Academies of Sciences, Engineering, and Medicine. 2015. *Trust and Confidence at the Interfaces of the Life Sciences and Society: Does the Public Trust Science? A Workshop Summary.* Washington, DC: National Academies Press, https://www.nap.edu/download/21798#. Accessed 12 July 2018.

More than 10 years ago, it had been noted that the majority of published research findings in the biomedical area were false.[1] This included epidemiological studies which have played a major role in several of the scientific stories related in this book — dietary fat, vaccination and GMO foods — as well as clinical trials, meta-analyses and cutting-edge research in molecular biology. By 2012, the situation had become even worse. Scientists at biotechnology firm Amgen in California were able to confirm only 11% of published landmark studies in their own field of cancer biology: an astonishing 89% of the studies were not reproducible, they found. Around the same time, a team at Bayer in Germany reported that only about 25% of published preclinical studies on potential new drugs could be validated. Some of the unreproducible papers had spawned completely new fields of research, generating hundreds of secondary publications. More worryingly, other papers had led to clinical trials that weren't likely to be of any benefit to the participants.[2]

Although there isn't much data on irreproducibility in fields outside medicine and biology, such as climate science or geology, a 2016 *Nature* survey revealed that none of the major areas of science are entirely immune to the problem — the highest rates of reproducibility being in physics and chemistry. The same survey found that over 70% of individual researchers have tried unsuccessfully to reproduce another scientist's results, and more than 50% have been unable to replicate even their own experiments.[3] In psychology, an attempt to replicate 100 reported studies found that even in the 39% that were reproducible, the size of the measured effect was reduced by about half compared with the original study.[4]

The reproducibility crisis is the topic of a recent report by the U.S. National Association of Scholars,[5] as well as a book by U.S. National Public Radio science correspondent Richard Harris.[6] At the outset, Harris pins the blame for the sorry state of current medical research on scientists taking shortcuts around the once hallowed scientific method. He cites numerous instances of the breakdown of the scientific process in biomedical

[1] Ioannidis, John P. A. 2005. "Why Most Published Research Findings are False." *PLoS Medicine*, 2(8): e124, 1-6.

[2] Begley, C. Glenn and Lee M. Ellis. 2012. "Raise Standards for Preclinical Cancer Research." *Nature*, 483, 531-533.

[3] Baker, Monya. 2016. "Is There a Reproducibility Crisis?" *Nature*, 533, 452-454.

[4] Open Science Collaboration. 2015. "Estimating the Reproducibility of Psychological Science." *Science*, 349, 943-951.

[5] Randall, David and Christopher Welser. 2018. *The Irreproducibility Crisis of Modern Science: Causes, Consequences, and the Road to Reform*, April. New York: National Association of Scholars, https://www.nas.org/projects/irreproducibility_report. Accessed 12 July 2018.

[6] Harris, Richard. 2017. *Rigor Mortis: How Sloppy Science Creates Worthless Cures, Crushes Hope, and Wastes Billions*. New York: Basic Books.

experiments. One disturbing example is research on breast cancer that was carried out on misidentified skin cancer cells, resulting in thousands of papers published in prominent medical journals on the wrong cancer. Harris also attributes the lack of progress in conquering disease — only 500 out of 7,000 known diseases being treatable today — to the absence of rigor in contemporary medical research.

But beyond the corrosive effect of irreproducibility, science is being attacked even more forcefully by rising corruption. I've discussed several examples of corrupted science in the book: the two most blatant cases both involved the discredited claims of anti-vaccinationists that vaccines can cause brain damage or autism in children. In the 1987–88 UK trial on alleged side effects of the whooping cough vaccine, the trial judge chastised neurologist Wilson and public health physician Miller for lying in their two studies and manipulating the gruesome details of neurological illnesses in sick children to falsely link the children's' disorders to the DTP vaccine. In the more egregious episode that claimed to connect autism to the measles vaccine, we saw in Chapter 6 how gastroenterologist Wakefield had his paper in *The Lancet* first retracted by the journal's editors, then declared fraudulent because the editors had unearthed evidence that he had falsified his data. His fraud was compounded by ethical lapses and medical misconduct surrounding the same study, resulting in Wakefield being disgraced and his medical license revoked. More recently, in 2018 *The Lancet* retracted two papers by thoracic surgeon Paolo Macchiarini, who was also disgraced following an investigation by the Karolinska Institute in Sweden that found him guilty of medical misconduct.[1]

Other instances of scientific malfeasance that border on corruption are the USDA's publication of its *Dietary Guidelines* over the solid scientific objections of its own review panel and the NAS Food and Nutrition Board; the Climategate emails and ongoing efforts to suppress contrary opinions in the global warming debate; and Quist and Chapela's astounding claim that CaMV 35S promoter genes had hopped between Mexican cornfields, an assertion that led *Nature* to publicly withdraw its support for the paper the two authors had already published in the journal.

While not yet at the same level as unreproducible research, outright fraud is also becoming a problem in science. Fraud goes much further than the lack of sufficient rigor and sometimes questionable ethics behind the reproducibility crisis: it involves falsifying or even making up data. A 2012 study published by the NAS discovered that the percentage of articles in

---

[1] Editorial. 2018. "The Final Verdict on Paolo Macchiarini: Guilty of Misconduct." *The Lancet*, 392, 2.

the biomedical research area retracted due to fraud had increased almost 10 times since 1975. In addition to fraud, a smaller percentage of articles had recently been retracted because of plagiarism or duplicate publication. Although the current percentage retracted due to fraud was still very small, the study authors pointed out that it underestimated the actual percentage of fraudulent articles in the literature, since only a fraction of fraudulent articles are retracted.[1]

A typical example is the so-called Indo-Mediterranean Diet Heart Study. This was a series of randomized controlled trials and other studies on the effects of a Mediterranean-style diet on coronary heart disease, directed by Indian cardiology researcher Ram Singh. His first paper, published in the *BMJ* in 1992,[2] soon aroused the suspicions of the journal's editor when reviewers of subsequent papers submitted for publication discovered discrepancies in the data. An analysis commissioned by the *BMJ* of one of the submitted papers concluded that Singh's dietary data was most likely fabricated, although the *BMJ*'s own investigation wasn't able to actually prove fraud.[3] Irregularities were also uncovered in a later Singh paper published by *The Lancet* in 2002[4] although, perhaps conveniently, the study consent forms and protocol, which contained details of the dietary intervention, had been eaten by termites. *The Lancet* was troubled enough about the paper to issue an official "Expression of Concern."[5] Nevertheless, despite the extremely strong suspicions of fraud by both journals, neither paper has been retracted to this day.

Fraud in biomedical research shows no sign of letting up, even though several recent high-profile cases have gone beyond retraction and scientific rebuke, resulting in criminal charges against the perpetrators who have sometimes been jailed. In 2010, Massachusetts anesthesiologist Scott Reuben

[1] Fang, Ferric C., R. Grant Steen and Arturo Casadevall. 2012. "Misconduct Accounts for the Majority of Retracted Scientific Publications." *Proceedings of the National Academy of Sciences*, 109, 17028-17033; Fang, Ferric C., R. Grant Steen and Arturo Casadevall. 2013. Correction for "Misconduct Accounts for the Majority of Retracted Scientific Publications." *Proceedings of the National Academy of Sciences*, 110, 1137.
[2] Singh, Ram B., Shanti S. Rastogi, Rakesh Verma et al. 1992. "Randomised Controlled Trial of Cardioprotective Diet in Patients with Recent Acute Myocardial Infarction: Results of One Year Follow Up." *BMJ*, 304, 1015-1019.
[3] Smith, Jane and Fiona Godlee. 2005. "Investigating Allegations of Scientific Misconduct: Journals Can Do Only So Much; Institutions Need to be Willing to Investigate." *BMJ*, 331, 245-246.
[4] Singh, Ram B., Gal Dubnov, Mohammad A. Niaz et al. 2002. "Effect of an Indo-Mediterranean Diet on Progression of Coronary Artery Disease in High Risk Patients (Indo-Mediterranean Diet Heart Study): A Randomised Single-Blind Trial." *The Lancet*, 360, 1455-1461.
[5] Horton, Richard. 2005. "Expression of Concern: Indo-Mediterranean Diet Heart Study." *The Lancet*, 366, 354-356.

was sentenced to six months in jail for fabricating data about clinical trials on the painkillers Vioxx, Celebrex and Bextra, often prescribed at the time to lessen postoperative pain after knee and other orthopedic surgeries. Reuben was paid hundreds of thousands of dollars by the drug manufacturers to promote their products, cash that he was ordered by the court to pay back. Over a 15-year period he published 21 fabricated papers, 13 of which were later retracted; some studies included false co-authors and imaginary patients.[1] In 2015, Iowa State University researcher Dong Pyou Han was sentenced to more than four and a half years in prison and ordered to repay $7.2 million in grant funds, after being convicted of fabricating and falsifying data in trials of a potential HIV vaccine. Han had spiked blood samples from vaccinated rabbits with human HIV antibodies, to make it appear that the vaccine enhanced immunity against HIV.[2] Both the Han and Reuben cases echo the earlier Wakefield episode mentioned above. Numerous other cases of scientific fraud have been reported as well, involving researchers outside the U.S. and in other fields.[3]

We'll examine the reasons underlying the reproducibility crisis and the rise in scientific fraud shortly. But in the light of these assaults on science, it should be little wonder that what we might label "fake science" in today's vernacular is so prevalent. By fake science I don't mean pseudoscience, which pretends to be science but is instead based on faith in some belief. Rather, fake science refers to the often ill-conceived studies that catch media attention, mostly in the nutritional area, proclaiming the health benefits or perils of some foodstuff or beverage. Coffee, for example, once extolled as a cure for everything from indigestion to miscarriages, in the 1970s and 1980s became villainized as a cause of heart attacks, only to have that association reversed in the last five years and to be seen currently as lowering the risk of heart disease and extending longevity, among other bonuses.[4] A steady stream of flip-flops like this doesn't help the credibility of science.

[1] Edwards, Jim. 2010. "Doc Who Faked Pfizer Studies Gets 6 Months in Prison, Showing Why Gift Bans are a Good Idea." *CBS MoneyWatch*, 25 June, http://www.cbsnews.com/news/doc-who-faked-pfizer-studies-gets-6-months-in-prison-showing-why-gift-bans-are-a-good-idea/. Accessed 12 July 2018.

[2] Leys, Tony. 2015. "Ex-Scientist Sentenced to Prison for Academic Fraud." *USA Today*, 1 July, https://www.usatoday.com/story/news/nation/2015/07/01/ex-scientist-sentenced-prison-academic-fraud/29596271/. Accessed 12 July 2018.

[3] Freckleton, Ian. 2016. "When Scientists Lie." *ABC: Ockham's Razor*, 26 July, http://www.abc.net.au/radionational/programs/ockhamsrazor/when-scientists-lie-ian-freckleton-on-scholarly-misconduct-fraud/7660464#. Accessed 12 July 2018.

[4] LaMotte, Sandee. 2018. "Health Effects of Coffee: Where Do We Stand?" *CNN Health*, 12 April, http://www.cnn.com/2015/08/14/health/coffee-health/. Accessed 12 July 2018.

The Insidious Influence Of Politics

So far in this chapter, we've examined forces attacking science from within — the lack of proper attention to empirical evidence, insufficient rigor in scientific research, dubious ethics, and the falsification and fabrication of data. But outside forces, dominated by politics, are contributing just as much if not more to the attack.

In the modern world, science is probably inseparable from politics because science plays such an important role in almost everything we do. Indeed, political forces can be seen at work in all of the book's illustrative examples: politics have been intricately interwoven with the debates over dietary fat, climate change and GMO foods, and have been a significant, if lesser part of the tussle between evolution and creationism, and of the vaccination debate. The reason that politics have intruded more into the dietary fat, climate change and GMO food debates is that the science in all three of these areas is still at the formative stage and so more vulnerable to external influences. In the case of creationism, however, the science of evolution was becoming well established when five U.S. states decided in the 1920s to enact laws banning or curtailing the teaching of evolution in public schools, and later when efforts were made by states to legislate teaching of creationism. Similarly, the efficacy of vaccination was broadly accepted long before the modern anti-vaccination movement in the U.S. began to campaign against mandatory school vaccination laws introduced in the 1960s and 1970s.

The incursion of politics into nutrition science began with the AHA's Nutrition Committee, to which Keys was appointed in 1961. Keys saw to it that the committee's report on his cherished dietary fat hypothesis, issued that same year, endorsed the now discredited connection between heart disease and saturated fat. Reinforced by the AHA's imprimatur, and by the close ties between the AHA and government agencies such as the NIH and CDC, Keys' hypothesis became the mainstay of *Dietary Goals* published by the U.S. government in 1977. The institutionalized bias inherent in *Dietary Goals* and the subsequent *Dietary Guidelines* was propagated for many years afterward by rivalries and political infighting between various branches of government such as the USDA, NIH and NHLBI. Only recently has nutritional science begun to regain the ground taken over by politics, paving the way to overturning the diet–heart hypothesis.

The entanglement of climate science with politics has similar origins, dating back to the ice age panic of the 1970s and the simultaneous emergence of the environmental movement. Not long after the global cooling spell that had commenced in 1940 ended around 1970 and the earth started warming again, the Charney Report endorsed the CO_2 hypothesis. The hypothesis, linking increased temperatures to steadily rising levels of CO_2 and other greenhouse gases, mostly from human emissions, underpins the belief in dangerous man-made global warming. Just as Keys' diet–heart hypothesis was embraced by the AHA and thus entered the political arena, the government-commissioned Charney Report thrust the CO_2 hypothesis into the political world, this time at the international level. Fewer than 10 years later the IPCC was founded. But, while the IPCC professes to be a purely scientific organization producing regular reports on climate change, in reality it mixes science with politics almost constantly, and is one of the leading advocates of the theory that humans are to blame for most of recent global warming. With almost 200 parties to the Paris Agreement and leading global corporations supporting the IPCC position, despite a paucity of empirical evidence for the theory, it's clear that politics have taken over from science.

Keys' clout in nutritional politics propped up his dietary fat hypothesis far longer than it should have survived, despite ever-mounting evidence from both epidemiological studies and major clinical trials on CHD and dietary fat that conflicted with the hypothesis. Likewise, the IPCC and environmental activists have kept the CO_2 hypothesis alive in spite of observational evidence that is either weak or doesn't match the predictions of the hypothesis at all. The result has been the abandonment of reason and the subversion of science to politics, fueled by fear and a quasi-religious belief in the narrative of catastrophic anthropogenic climate change. When the narrative is used to close off debate on global warming, science is unquestionably under assault.

The tendency toward hype in politics also subverts science. The select U.S. government committee that drew up the new low-fat dietary recommendations in 1977, fighting for its political future in Congress, wildly exaggerated its message by declaring that excessive fat or sugar in the diet was as much of a health threat as smoking, even though a reasoned evaluation of the evidence showed such a claim to be untrue. Not long afterwards, scientists themselves succumbed to the same tactics in the public debate over "nuclear winter" — a potentially catastrophic global chilling that would result from a nuclear war. A 1983 paper famously predicted that a widespread nuclear exchange would encase the planet in vast clouds of smoke and dust from burning cities, blocking out sunlight.[1] Ironically, one of the paper's authors was American astrophysicist Carl Sagan (1934–1996),

---

[1] Turco, R. P., O. B. Toon, T. P. Ackerman, J. B. Pollack and Carl Sagan. 1983. "Nuclear Winter: Global Consequences of Multiple Nuclear Explosions." *Science*, 222, 1283-1292.

popularizer of science and host of the award-winning TV series *Cosmos*. But, although Sagan was a strong advocate of the scientific method, having said "Science invites us to let the facts in, even when they don't conform to our preconceptions,"[1] he was not above sensationalizing the effects of nuclear war for political purposes. By 1990, however, the five authors of the paper conceded that they had misinterpreted the evidence and overestimated the severity of nuclear winter.[2]

The more recent slew of blatant exaggerations in climate science for political purposes — including the inflated predictions of climate models, the exaggerated consensus among climatologists on human-induced warming, and supposed shrinking of the polar bear population — border on fraud and are an abuse of science. A similar pattern has emerged in the science of GMO foods, with Ho dramatically overstating the possibility of dormant viruses or cancer genes being activated by the genetic engineering process, and Quist and Chapela claiming to have discovered wandering transgenes in Mexican corn. Such exaggeration does science a gross disservice.

Politics is more visceral than rational and, while political agendas on scientific issues may purport to be about science, only rarely do the evidence and logic intrinsic to science play a big role. A commonly exploited emotion is fear. Fear of disastrous consequences of global warming was a major factor behind the world's enactment of the UNFCCC, the Kyoto Protocol and the Paris Agreement, even though the actual evidence for a looming catastrophe is nonexistent. And irrational fears about the safety and ecological effects of genetic engineering have led to the wholesale rejection of agricultural science by a major portion of the public, and delayed the development of potentially lifesaving new crops such as golden rice.

The inseparability of science and politics has given rise to the concept of postnormal science, which I discussed briefly at the end of Chapter 5. Postnormal science, a term derived from Kuhn's definition of "normal science,"[3] is an attempt to manage areas of science in the public sphere where the evidence is incomplete and highly uncertain, generating disputes over its interpretation.[4] The concept is closely linked to the Precautionary Principle that, as we've already seen, underlies the UNFCCC, as well as the bans on

[1] Sagan, Carl. 1997. *The Demon-Haunted World: Science as a Candle in the Dark*. New York, Ballantine Books, 27.

[2] Browne, Malcolm W. 1990. "Nuclear Winter Theorists Pull Back." *New York Times*, 23 January, http://www.nytimes.com/1990/01/23/science/nuclear-winter-theorists-pull-back.html?pagewanted=all. Accessed 12 July 2018.

[3] Kuhn, Thomas S. 2012. *The Structure of Scientific Revolutions*. Chicago: University of Chicago Press.

[4] Funtowicz, Silvio O. and Jerome R. Ravetz. 1993. "Science for the Post-Normal Age." *Futures*, 25, 739-755.

GMO crops by more than half the countries in the European Union. But in trying to bridge the gap between science and politics, postnormal science only clouds the picture. One of its stated intentions is to bring scientific issues "into public and official consciousness by campaigns involving activists and the media," by means of "extended peer communities" who resort to unscientific sources of information such as anecdotal evidence and leaked documents.[1] The advocacy and bias of this approach encourages the subjugation of truth and reason — the very essence of traditional science.

The intrusion of politics into science dates back to the period after World War II, when the escalating costs of large science projects resulted in a sudden jump in government funding of science. As noted at the end of Chapter 1, public funding creates a conflict of interest between the remuneration and integrity of individual scientists. In 1971, British mathematician and biologist Jacob Bronowski (1908–74), author of the acclaimed TV documentary *The Ascent of Man*, echoed Eisenhower's farewell address in expressing concern about the integrity of science as a whole:

> We see the growing involvement year by year of government in science and science in government; and unless we cut that entanglement, we endanger the integrity of all science, and undermine the public trust in it.... The silent pressure for conformity exists whenever grants and contracts for research are under the direct control of governments; and then ... no science is immune to the infection of politics and the corruption of power.[2]

A speaker at the 2015 NAS workshop on the life sciences and society concluded that the integrity of science had indeed lost the historical trust of the public, saying that "the American public, which generally has an unfavorable view of politicians, can extend its negative feelings toward science when it perceives that science is being used for political purposes." Another participant commented that "when there is societal debate, public trust often becomes a function more of political ideology than of scientific fact."[3] And in a journal article, a policy expert has remarked that scientists need to cultivate "the self-awareness to recognize and openly acknowledge the relationship between their political convictions and how they assess scientific evidence."[4]

---

[1] Ravetz, Jerry. 2004. "The Post-Normal Science of Precaution." *Futures*, 36, 347-357.

[2] Bronowski, J. 1971. "The Disestablishment of Science." In Watson Fuller, ed., *The Social Impact of Modern Biology*, London: Routledge & Kegan Paul, chap. 19.

[3] U.S. National Academies of Sciences, Engineering, and Medicine. 2015. *Trust and Confidence at the Interfaces of the Life Sciences and Society: Does the Public Trust Science? A Workshop Summary.* Washington, DC: National Academies Press, 21, https://www.nap.edu/download/21798#. Accessed 12 July 2018.

[4] Sarewitz, Daniel. 2015. "Reproducibility Will Not Cure What Ails Science." *Nature*, 525, 159.

One way of achieving this and avoiding the inherent conflicts between science and politics is the use of "red teams," recently suggested as a means of questioning the conventional climate change narrative.[1] As I discussed before, the climate change narrative has been used to suppress both contrary views on the origins of global warming and even the evidence itself. The reasons behind the suppression include confirmation bias, breakdown of the peer review process, and funding mechanisms for climate change research, all of which we'll look at shortly. A red team is an independent group that challenges an organization's accepted wisdom by assuming a devil's advocate role, in order to examine the thinking and probe the intentions of both the organization and potential rivals. The practice has been widely utilized by the military, intelligence agencies and industry to identify vulnerabilities or size up the competition.[2] Its extension to scientific policy making could well be beneficial.

Why Is Science Being Assailed?

With science under assault from so many different directions, we might well wonder why — why is scientific evidence being misrepresented or ignored so often, why is science becoming more corrupt and susceptible to political influence? The answers to these questions are many. Some of the answers we've considered already; the others have been the subject of extended debate, especially on the editorial pages of scientific journals and on the Internet.

In the biomedical arena, where researchers literally encounter issues of life and death, if only tangentially, the epidemic of irreproducibility and the rising incidence of fraud can be tied directly to pressures experienced by today's scientists. Prominent among these pressures is the perceived need for scientists to publish their work in high-impact journals. Journal editors vie for the prestige of the highest impact factor, which measures how often on average articles published in a journal are cited by other authors. The splashier a paper submitted for publication, therefore, the more likely it is to be accepted — but often at the cost of research quality. The temptations are strong for an author to embellish results or mold the data to fit an underlying hypothesis, instead of modifying the hypothesis to fit the data

[1] Harvey, Chelsea. 2017. "These Scientists Want to Create 'Red Teams' to Challenge Climate Research. Congress is Listening." *Washington Post*, 29 March, https://www.washingtonpost.com/news/energy-environment/wp/2017/03/29/these-climate-doubters-want-to-create-a-red-team-to-challenge-climate-science/?utm_term=.d8be29e1a8f7. Accessed 12 July 2018.

[2] Zenko, Micah. 2015. *Red Team: How to Succeed by Thinking Like the Enemy*. New York: Basic Books.

as the scientific method demands. The conflict of interest between scientific integrity and career advancement, signified by promotion or tenure as well as other forms of recognition, can make it more important to be the first to publish or to present sensational findings than to be correct. And the need to secure ongoing research funding only adds to the pressure to report results even when they're uncertain or unreliable.

The emphasis on publishing research reports in journals with high impact factors and on winning research grants is highly detrimental to science, because it encourages scientists to pursue short-term success by cutting corners instead of conducting careful, well thought-out studies that take much longer to produce less spectacular, but more robust, findings. Rather than challenging authority and the status quo as effective researchers should, too many scientists today sacrifice their integrity for the enticing appeal of success, power and monetary reward. This is true across the scientific spectrum, not just in biomedicine. And the lack of integrity is exacerbated by abuses of the peer review process, once an effective mechanism for uncovering shortcomings and inaccuracies in scientific papers, but now as corrupted as research itself.

A development that mocks peer review is the rise of so-called predatory open access journals that publish scientific papers with minimal if any peer review at all. Open access journals date from the advent of the Internet and provide online papers free to readers, unlike conventional online journals that often charge hefty fees. Instead, open access publishers, many of which are trustworthy, usually charge authors a fee to cover the cost of peer review, editing and website maintenance. Predatory publishers, however, exploit this model by charging a fee without offering the same publication services. Because of the limited or total lack of peer review, predatory open access journals are susceptible to corruption by accepting without question papers that a reputable journal would turn down. As an illustration of their dishonesty and lack of rigor, a recent sting investigation found that a staggering 33% of predatory journals contacted offered a sham scientific editor a position on their editorial boards, four of them immediately appointing the fake scientist as editor-in-chief. The fictitious editor listed no published papers in her résumé nor any reviewing or editorial experience.[1]

One of the most common weaknesses of contemporary scientific research, and one that has contributed heavily to the reproducibility crisis, is the misunderstanding and misuse of statistics. The statistical process of regression analysis, such as the least-squares fitting method familiar to

---

[1] Sorokowski, Piotr, Emanuel Kulczycki, Agnieszka Sorokowska and Katarzyna Pisanski. 2017. "Predatory Journals Recruit Fake Editor." *Nature*, 543, 481-483.

most scientists, is used to quantify relationships between variable factors in a study. But regression analysis is often misused to prove causation, even though the process can only determine association, not cause and effect. Another popular but faulty practice is data dredging,[1] which refers to the use of the same data that shows a pattern or apparent correlation to demonstrate the statistical significance of the pattern or correlation; independent data must be used to verify any correlation. Such phenomena are not new, however. It was over 100 years ago that suspicion of statistical practices induced American author Mark Twain to make his famous remark that there are three kinds of lies: lies, damned lies and statistics.[2]

According to Greek medical researcher John Ioannidis, well known for his examination of bias in the biomedical literature, the inability to replicate scientific discoveries often stems from the "ill-founded strategy" of drawing definitive conclusions from a single study when the statistical p-value is less than 0.05.[3] The smaller the p-value, the more likely it is that the experimental data can't be explained by existing theory and that a new hypothesis is needed;[3] p-values below 0.05 are commonly regarded as statistically significant. However, despite this belief, a p-value on its own doesn't provide evidence for or against a particular hypothesis.[4] Yet the medical community adheres to the mistaken notion that p-values are essential to the interpretation of research results. In addition, Ioannidis has found that research findings are much more likely to be true, and less likely to be plagued by false positives, in large randomized controlled trials and meta-analyses than in smaller randomized trials or in epidemiological studies. But as discussed in Chapter 4, even meta-analyses and systematic reviews can produce conflicting results.

In this book we've also come across two notable examples of statistical abuse in other areas of science. The first was Mann's hockey stick in the climate change debate — the infamous attempt to distort the historical temperature record, in order to bolster the IPCC's assertion that humans are to blame for the bulk of modern global warming. Many statisticians have faulted Mann's statistical analysis, a U.S. National Research Council study noting that even completely random data can sometimes generate a hockey

---

[1] Begley, C. Glenn, Alastair M. Buchan and Ulrich Dirnagl. 2015. "Institutions Must Do Their Part for Reproducibility." *Nature*, 525, 25-27.

[2] 2018. Directory of Mark Twain's Maxims, Quotations, and Various Opinions: Statistics, http://www.twainquotes.com/Statistics.html. Accessed 12 July 2018.

[3] Or the less likely that the data can be explained by the null hypothesis.

[4] Wasserstein, Ronald L. and Nicole A. Lazar. 2016. "The ASA's Statement on p-Values: Context, Process, and Purpose." *The American Statistician*, 70, 129-133.

stick shape.[1] The second instance of misused statistics, in this case based on overconfidence in p-values, was in the vaccination story, when Miller attempted to claim that neurological illnesses he'd seen in children were linked to the DTP vaccine. As we saw in Chapter 6, Miller was excoriated by a trial judge for misrepresentation of his case histories, a deception that grossly distorted the statistical significance of his data.

With the increasing politicization of science, confirmation bias is becoming more and more of a problem. The traditional objectivity of science has been abandoned for outright advocacy in many fields. We should realize, however, that confirmation bias — the selective evaluation of evidence according to one's preconceptions or existing beliefs, and closely related to narrow-mindedness — is as old as science itself, and has been exhibited by many of the greatest scientific minds over the ages. Indeed, some psychologists have argued that confirmation bias has played a stabilizing role in science, by guarding against uncritical adoption of new ideas before they have been adequately tested, and against throwing out old theories too hastily.[2] But sometimes confirmation bias works to unnecessarily delay acceptance of new theories. It was confirmation bias on the part of geophysicists Cox and Doell, who cherry-picked their magnetic data to favor a wandering north pole over continental drift, that contributed to the initial rejection of Wegener's novel theory. Confirmation bias also impeded acceptance of Darwin's radical theory of evolution, allowing 19th-century scientists and the general public to cling to older creationist beliefs before genetics came on the scene.

The reverse is true in the case of dietary fat, where confirmation bias that began with Keys' single-minded focus on saturated fat and heart disease helped to shore up his diet–heart hypothesis almost from the beginning. Similarly, in the case of climate change, confirmation bias underlies the belief in dangerous man-made global warming and the false proclamation that the debate is already settled. In the debates over vaccination and GMO foods, confirmation bias can be readily seen in the positions of anti-vaccinationists and anti-GMO activists: in the falsification of data on autism and the MMR vaccine by Wakefield, and in the flawed claims that GMOs are harmful by Pusztai and others.

The human tendency to treat evidence selectively stems in part from our tribal origins. Tens of thousands of years ago when humans roamed the earth

---

[1] U.S. National Research Council, Board on Atmospheric Sciences and Climate. 2006. *Surface Temperature Reconstructions for the Last 2,000 Years*. Washington, DC: National Academies Press, chaps. 9 and 11, http://www.nap.edu/catalog/11676/surface-temperature-reconstructions-for-the-last-2000-years. Accessed 12 July 2018.

[2] Nickerson, Raymond S. 1998. "Confirmation Bias: A Ubiquitous Phenomenon in Many Guises." *Review of General Psychology*, 2, 175-220.

as hunter-gatherers, the irrational beliefs and groupthink of tribalism were necessary for survival, both physically in warding off predators and socially in maintaining tribal cohesion. Loyalty to the tribe and conformity were more highly valued than dissent or skepticism.[1] But despite our transition from a tribal existence to farming life and civilization, we're not really that different today. Tribal behavior was very much on display in the denigration of rock magnetism by geologists who refused to accept the solid evidence for continental drift; in the biased reaction of Keys believers to meta-analyses and systematic reviews that showed no link between saturated fat and CHD events, and to Teicholz's *BMJ* article questioning the science behind *Dietary Guidelines*; and in the circling of the wagons by climate change warmists, following the Climategate revelations of their attempts to manipulate data and suppress evidence.[2] The more threatened we feel, the more we turn to the tribe for safety.

Tribalism also manifests itself in *ad hominem* attacks, which are a defensive measure against ideas that threaten tribal unity. Of the attacks described in this book, the most vicious were undoubtedly those directed at Wegener while he was alive — abuse that forced the young professor to leave his native Germany to find a tenured university position. But, while not as mean-spirited as the bare-knuckle political rebukes typical of the 19th and early 20th centuries, the *ad hominem* attacks on Wegener pale in comparison with the death threats that accompany the 21st century debates on global warming and vaccination. Unfortunately, though the vast majority of us are no longer hunter-gatherers, tribalism is still part of human nature. The increased prevalence of tribal behavior in modern science compared with the past most likely reflects intensified competition among an ever growing body of scientists.

The tribal instinct of groupthink,[3] which values tribal harmony over critical evaluation of the evidence, encourages agreement with the consensus. Some scientists argue that consensus has no place in science, that the scientific method alone with its emphasis on reproducible results dictates whether a particular hypothesis stands or falls. But the eventual elevation of a hypothesis to a theory, such as the theory of evolution or the theory of plate tectonics, does depend on a consensus being reached among the scientific

---

[1] Ronfeldt, David. 2006. "In Search of How Societies Work: Tribes — The First and Forever Form," Working Paper WR-433-RPC, December. Santa Monica, CA: RAND Pardee Center.

[2] Curry, Judith. 2014. "The Legacy of Climategate: 5 Years Later." *Climate Etc.* blog, 1 December, https://judithcurry.com/2014/12/01/the-legacy-of-climategate-5-years-later/. Accessed 12 July 2018.

[3] Psychology Today. 2018. "Groupthink," https://www.psychologytoday.com/basics/groupthink. Accessed 12 July 2018.

community. Nonetheless, the consensus, often reinforced by groupthink, isn't always backed by the evidence and so isn't always right.

In fact, as we've seen in earlier chapters, both an established scientific consensus and a newly formed one can be wrong. This was certainly the case for Ptolemy's geocentric theory of the solar system that Galileo fought to overturn, advocating instead the rival heliocentric theory of Copernicus. The same was true too for the contraction theory of a cooling earth that Wegener and, later, Holmes battled in championing the revolutionary continental drift theory; as well as for the entrenched consensus on creationism in the 19th and early 20th centuries, before Darwin's theory of evolution gained acceptance. And despite the broad initial consensus on Keys' dietary fat hypothesis, that has also turned out to be faulty. Even the consensus that climate change is largely a consequence of human activity — a consensus that falls far short of the purported 97% of climate scientists, as discussed in Chapter 5 — is on very shaky ground, because of its dependence on theoretical computer climate models instead of actual observations. The only illustrative examples in the book in which the initial consensus has endured so far, although less strongly among the general public than among scientists, are vaccination and GMO foods.

It may be interesting for readers to contemplate the connection, if any, between the present consensus on each of our six topics and the scientific evidence. The conventional wisdom is that science backs the consensus on continental drift, evolution, climate change, vaccination and GMOs. Nevertheless, as our examination of each of these illustrative examples has demonstrated, science is on the side of those who accept the theory of evolution and the safety of vaccination and GMO foods, but science also supports those who question the dietary fat hypothesis and those who doubt the conventional wisdom on climate change. Dietary fat is in a category of its own, since the 50-year-old consensus linking deaths from CHD to saturated fat in the diet is currently being overturned. Creationism, the anti-vaccination and anti-GMO movements, and the notion of catastrophic global warming caused by humans are all based on faith, not science.

A popular belief is that skepticism about climate change is closely linked to skepticism about evolution, especially by religious conservatives, as part of a supposed anti-science trend. In the U.S., 34% of the population are creationists who reject evolution entirely,[1] and about the same percentage

---

[1] Pew Research Center. 2015. "U.S. Public Becoming Less Religious. Chapter 4: Social and Political Attitudes," 3 November, http://www.pewforum.org/2015/11/03/chapter-4-social-and-political-attitudes/. Accessed 12 July 2018.

are skeptics on climate change.[1] But a recent study has found that only about 65% of American anti-evolutionists are climate change skeptics: the remaining 35% who reject evolution believe in the climate change narrative.[2] So the two groups of skeptics don't consist of the same individuals, although there is some overlap. In the case of GMO foods, again approximately equal percentages of the public reject the consensus among scientists that GMOs are safe to eat, and are skeptical about climate change.[3] Once more, however, the two groups don't consist of the same people. And, although most U.S. farmers accept the consensus on the safety of GMO crops but are climate change skeptics, [4] there are environmentalists who are GMO skeptics but accept the prevailing belief on climate change; Prince Charles is a notable example of the latter.

Whither Science?

Modern science is under attack by multiple forces, including both the attitudes and behavior of individuals, and organized groups of people such as creationists, anti-vaccinationists and anti-GMO protestors. While the 21st century has seen astonishing scientific advances in many fields such as space exploration, genetics and synthetic materials, the laboriously constructed edifice of the scientific method is showing cracks. Overall, what we call science today is not the science that led Newton to the theory of gravitation, or drew Einstein to his theory of relativity, or uncovered the structure of DNA, or put a man on the moon. Current science is beset by problems ranging from mistreatment of evidence to irreproducibility to fraud to political interference. In bemoaning the reproducibility crisis in the biomedical sciences, *The Lancet*'s editor-in-chief Richard Horton has written:

> The case against science is straightforward: much of the scientific literature, perhaps half, may simply be untrue.... [S]cience has taken a turn toward darkness.[5]

[1] Swift, Art. 2017. "In U.S., Belief in Creationist View of Humans at New Low." *Gallup*, 22 May, https://news.gallup.com/poll/210956/belief-creationist-view-humans-new-low.aspx. Accessed 12 July 2018.

[2] Howard Ecklund, Elaine, Christopher P. Scheitle, Jared Peifer and Daniel Bolger. 2016. "Examining Links Between Religion, Evolution Views, and Climate Change Skepticism." *Environment and Behavior*, online publication, 26 October, doi:org/10.1177/0013916516674246, table 2.

[3] McFadden, Brandon R. 2016. "Examining the Gap Between Science and Public Opinion About Genetically Modified Food and Global Warming." *PLoS One*, 11(11): e0166140, 1-14.

[4] Kowitt, Beth. 2016. "The Paradox of American Farmers and Climate Change." *Fortune*, 29 June, http://fortune.com/2016/06/29/monsanto-farmers-climate-change/. Accessed 12 July 2018.

[5] Horton, Richard. 2015. "Offline: What is Medicine's 5 Sigma?" *The Lancet*, 385, 1380.

As another commentator recently put it: "The problem with science is that so much of it simply isn't."[1] The growing lack of scientific integrity has contributed to the loss of trust in science, and to crises in public science such as the failure of government scientists in 2015 to detect dangerously high concentrations of lead in the Flint, Michigan water supply — which caused elevated blood-lead levels in thousands of children and resulted in criminal charges against state officials.[2]

Such a dismal picture raises the question, where is science headed? Is there any hope for the future? Science and the scientific method have made enormous progress since their humble beginnings in the academies of ancient Greece, but it's not at all clear that their lofty status, first achieved at the peak of the Scientific Revolution during the Age of Reason, is sustainable today. Evidence and logic, the twin hallmarks of science, are increasingly being ignored.

One thing that I've noticed myself is the dearth of comments in online articles about scientific fraud. With few exceptions, the number of comments is typically only a handful or at most a few tens, whereas articles on controversial topics, scientific or not, usually elicit hundreds of comments just in the first few days. When the number of responses to a piece on fraud is large, which is a rare event, the comments tend to be mostly from scientists. What this suggests is that few of those *outside* the scientific community care much about science. Although such observations are purely anecdotal, they reflect modern-day uncertainty about the *relevance* of science.

The relevance of science to society is a philosophical issue that has been discussed more and more often over the past 50 years or so. Feynman addressed the subject in a 1964 lecture titled "What Is and What Should Be the Role of Scientific Culture in Modern Society," ascribing the irrelevance of science to the general public at that time to ignorance about science and a lack of effective communication by scientists about the wonders of the discipline.[3] The subsequent Apollo program and the landing of the first human on the moon in 1969 briefly stirred interest in scientific activity. But the excitement had already subsided five years later, when Nobel-Prize winning British chemist Sir George Porter (1920–2002) again examined the question of relevance, proposing that the ultimate relevance of science is

[1] Wilson, William A. 2016. "Scientific Regress." *First Things* newsletter, May, https://www.firstthings.com/article/2016/05/scientific-regress. Accessed 12 July 2018.
[2] Kolowich, Steve. 2016. "The Water Next Time: Professor Who Helped Expose Crisis in Flint Says Public Science is Broken." *The Chronicle of Higher Education*, 2 February, http://www.chronicle.com/article/The-Water-Next-Time-Professor/235136#. Accessed 12 July 2018.
[3] Feynman, Richard P. 1999. *The Pleasure of Finding Things Out: The Best Short Works of Richard P. Feynman.* Cambridge, MA: Perseus Books, chap. 4.

to discover our purpose on Earth — our former religious sense of purpose having been diminished by scientific discoveries revealing the secrets of nature, as well as by rising secularism.[1]

The gap between scientists and the public on questions such as evolution, climate change and GMOs is another sign of the declining relevance of science, or the relevance of scientists at least. Scientists no longer have the credibility they once did, partly because of dwindling trust in science as mentioned before, and partly because of disillusionment over the failure of science so far to solve problems such as finding a cure for cancer. Recent surveys reported by the U.S. National Science Foundation (NSF) have found that about 50% of the population of most countries think that scientific knowledge is unimportant, and only 40% of the American public have "a great deal of confidence" in leaders of the scientific community, even though 70% of respondents believe that science does more good than harm.[2] Perhaps the glory days of science are over, a possibility that takes us back to the discussion in the opening chapter about science funding.

A common refrain in the scientific research community is that more funding is needed to abate fierce competition for the scarce funds that are available. It's widely believed that competition for research grants is one of the main underlying sources of the pressures on scientists, especially younger and less well-established ones, which are increasingly resulting in sloppy research, results that can't be replicated and unabashed fraud. Unfortunately, overall research funding is unlikely to increase in the future, at least in the U.S. and Europe. According to a 2013 report, U.S. federal expenditure on nondefense research and development has remained essentially the same at approximately 10% of the domestic discretionary budget for more than 40 years. And the percentage of successful research proposals funded by NIH and NSF programs had dropped from above 30% before 2001 to 20% or even lower in 2011.[3] From 2008 to 2015, the latest period for which data was available at the time of writing, federal government spending on research and development in the U.S. fell by about 7%, though this was amply compensated by a 17% rise in spending by industry,[4] which included almost doubled industry expenditure on basic research.[5]

[1] Porter, George. 1974. "The Relevance of Science." *Engineering and Science*, 38, 22-23.
[2] U.S. National Science Board. 2018. *Science and Engineering Indicators 2018*, NSB-2018-1. Alexandria, VA: National Science Foundation, chap. 7.
[3] Howard, Daniel J. and Frank N. Laird. 2013. "The New Normal in Funding University Science." *Issues in Science and Technology*, 30, 71-76.
[4] U.S. National Science Board, 2018, chap. 4.
[5] Mervis, Jeffrey. 2017. "Data Check: Federal Share of Basic Research Hits New Low." *Science*, 355, 1005.

But if funding for science has leveled off, and the number of scientists continues to escalate — some 80% to 90% of all the scientists who ever lived are thought to be alive today[1] — then the pressures experienced by researchers are only going to increase. Will science survive?

Personally, I'm optimistic that it will and that reason will return. During its long history, science has stagnated for an extended period more than once, in eras when the emphasis shifted from seeking out new knowledge to consolidation of existing knowledge. And even in periods when science moved forward, it endured periodic attacks. One of the most prolonged offensives was by the medieval church, whose official doctrine was that science should be subservient to theology, the policy challenged so boldly by Galileo. But, while Galileo himself was punished by the Inquisition and his *Dialogue* banned, the tome eventually disappeared from the Church's list of banned books, albeit not until two centuries later, vindicating both Galileo and science.

A more recent example is the setback of agriculture by more than 30 years in the 20th century Soviet Union, at the hands of biologist Trofim Lysenko (1898–1976), during a period when crop cultivation was making great strides elsewhere in the world. Lysenko, who became Stalin's director of agriculture and then genetics, falsely claimed that the mechanism for transforming winter wheat into spring wheat, by an old technique that he had successfully revived,[2] could be inherited. This claim, which is similar to the Lamarckian notion of acquired characteristics such as long giraffe necks (Chapter 2), flies in the face of Mendel's genetic theory. But Lysenko purged the Soviet scientific establishment of thousands of geneticists and other biologists — having them fired, banished to labor camps or even executed — and insisted that his methods would dramatically increase crop yields. In fact, harvests diminished, causing devastating famines. Lysenko himself was finally fired in 1965, after bringing Soviet agricultural, biological and genetics research to a standstill. Nevertheless, the Soviets gradually caught up to the rest of the world and again, science survived.[3]

A paradox is that at the same time modern science is ailing, technology — utilitarian applications of knowledge that today depend mostly on scientific

[1] de Solla Price, Derek J. 1963. *Little Science, Big Science*. New York: Columbia University Press, 1.

[2] The technique involved cold treating wheat seeds under high humidity, together with a new system of crop rotation. Lysenko also developed other questionable agricultural practices.

[3] Zielinski, Sarah. 2010. "When the Soviet Union Chose the Wrong Side on Genetics and Evolution." *Smithsonian Magazine*, 1 February, http://www.smithsonianmag.com/science-nature/when-the-soviet-union-chose-the-wrong-side-on-genetics-and-evolution-23179035/. Accessed 12 July 2018.

advances, rather than accumulated practical experience — is robust. Yet this isn't the first time in human history when science and technology have been divorced from each other. After blossoming during the Greek Golden Age, science withered for hundreds of years while technology thrived, as the Romans built tens of thousands of kilometers of sophisticated roads as well as numerous aqueducts and bridges. The cycle repeated itself once Rome fell, science briefly reviving in the medieval Islamic world before going into hibernation yet again until the advent of the Scientific Revolution centuries later. But while science faltered, technology progressed, with the development of the waterwheel, the printing press and new weaponry among other innovations. Following the Scientific Revolution, however, both science and technology advanced by leaps and bounds up to the present day, when science is under attack once more.

In trying to understand whether the current assault portends another, imminent "time-out" for science, should my own optimism not be warranted, it's worth reexamining possible reasons for the decline of ancient science. In Chapter 1, I discussed the prevailing view among science historians that Greek science came to a gradual halt because science never achieved a firm social footing, because "Greek society failed to assign a distinct role to the scientist," in the words of one historian.[1] But other reasons have also been advanced for the slide of Greek science into dormancy.

Prominent among these alternative explanations is the rise of religious sects and cults in antiquity. Cults in particular were a threat to the emergence of early science, its emphasis on empirical observation and reasoning being at odds with the astrological and occult beliefs of cultists. Science competed with cults, such as the cult of the Greek agriculture goddess Demeter, in attempting to explain the mysteries of nature. Fledgling Christianity too was often hostile to science, which was regarded by the early church as a pagan culture. However, while worship of Christ seems unlikely in itself to have caused the stagnation of Greek science, since ancient Greeks worshipped other gods as well, it's possible that wonder at the story of Christ supplanted the wonder that ancient science would have evoked. Alternatively, superstition and interest in the supernatural may have taken over, as they did later in the Middle Ages, or have done today in faith-based beliefs such as creationism and anti-vaccinationism.

From a moral standpoint, science writer and *Skeptic* magazine editor-in-chief Michael Shermer has argued that the rationality, empiricism and skepticism of science have been driving forces behind moral progress over

---

[1] Cohen, H. Floris. 1994. *The Scientific Revolution: A Historiographical Inquiry*. Chicago: University of Chicago Press, 257.

the last few centuries.[1] To my mind, such a claim is rather a stretch. On the other hand, renowned Yale University computer scientist David Gelernter is more pessimistic and believes that science is currently corrupted by its own power as much as the 16th-century Catholic Church was, and that any vestige of past humanism in science is in danger of being lost.[2]

But whatever the fate of science, there can be no doubt that it's currently experiencing a constant barrage of abuse and rejection. Even though the scientific method itself may be intact, fewer and fewer scientists are adhering to it. While the role of the time-honored scientific method in the era of big data has been questioned, because of the vast quantities of information generated, the importance of empirical evidence hasn't diminished — it's simply more difficult to uncover the evidence among massive volumes of data, a task for which statistical techniques are essential.[3] And logical analysis of the collected evidence is just as important today as it was in the time of Plato and Aristotle.

[1] Shermer, Michael. 2015. *The Moral Arc: How Science Makes Us Better People*. New York: Henry Holt and Company.

[2] Gelernter, David. 2018 (to be published). *Subjectivism: The Mind from Inside*. New York: W. W. Norton & Company.

[3] Swan, Melanie. 2015. "Philosophy of Big Data," slide presentation, https://www.slideshare.net/lablogga/philosophy-of-big-data. Accessed 12 July 2018.

GLOSSARY

ACLU	American Civil Liberties Union
ad hominem	a modern Latin phrase that means directed against the person, rather than their opinion or argument
alchemy	an occult precursor to modern chemistry, one of its main goals being to turn base metals such as lead into gold or silver
AHA	American Heart Association
ASCN	American Society for Clinical Nutrition
astrology	a pseudoscience that makes predictions of the future based on the apparent motion of stars across the sky; in the ancient world, however, astrology was considered to be a science, ranked just as highly as astronomy
Bt	a soil bacterium (*Bacillus thuringiensis*), proteins from which are widely used in organic pesticides for crops
CaMV 35S	a popular gene transfer promoter used in genetic engineering of plants
CDC	U.S. Centers for Disease Control and Prevention
Celsius	the metric temperature scale
Charney Report	a 1979 report commissioned by the NAS to critically assess early computer climate model predictions of future global warming
CHD	coronary heart disease
clinical trial	also known as a randomized controlled trial, a clinical

	trial is a study that investigates the effect of a health-related intervention by assigning participants at random to two identical groups, only one of which receives the intervention, with the other group used as a control
CRS	Creation Research Society
CSI	complex specified information, a concept in ID
Dietary Goals	a set of recommended dietary guidelines for the American public, issued in 1977 by the U.S. government and based on a study by the Senate Select Committee on Nutrition and Human Needs
Dietary Guidelines	the U.S. government's official dietary guidelines, issued at five-yearly intervals since 1980
DOE	U.S. Department of Energy
DTP	diphtheria, tetanus, and pertussis
El Niño	a natural climate cycle that causes temperature fluctuations and other climatic effects in tropical regions of the Pacific Ocean; El Niño is the warm phase of the El Niño — Southern Oscillation (ENSO)
EPA	U.S. Environmental Protection Agency
epidemiological study	
	also known as an observational study, an epidemiological study seeks to identify possible causes of disease or other health-related conditions in a population, but can only establish association, not causation
FDA	U.S. Food and Drug Administration
geocentric	earth-centered, referring to an early model of the solar system proposed by Ptolemy in the 2nd century CE
GISS	NASA Goddard Institute for Space Science
GMO	genetically modified organism
Golden Age	the historical era in ancient Greece from approximately 600 BCE until the death of Alexander the Great in 323 BCE, a period when Greek arts and science thrived
golden rice	rice genetically engineered to contain beta-carotene, a natural yellow pigment that produces vitamin A in the human body; one of the necessary genes comes from daffodils
HadCRU	the collaboration between the Climatic Research Unit at the University of East Anglia, UK and the UK Met

	Office's Hadley Centre
heliocentric	sun-centered, referring to our current model of the solar system, first proposed by Aristarchus in the 3rd century BCE and resurrected by Copernicus in 1543
ice age	a lengthy cold period lasting 10,000 years or more, during which massive ice sheets and glaciers covered large areas of the planet in the past and the average global temperature was about 6° Celsius (11° Fahrenheit) lower than today
ID	intelligent design, a form of creationism whose proponents believe that the natural world was created by an intelligent designer, who may or may not be God
IPCC	the UN's Intergovernmental Panel on Climate Change
Kyoto Protocol	a UN protocol that establishes mandatory limits for emissions of CO_2 and other greenhouse gases by industrialized countries, with the stated objective of stabilizing greenhouse gas concentrations in the atmosphere; the Protocol took effect in 2005 and has currently been extended to 2020
Lamarckism	a misnamed concept of biological inheritance promoted by Lamarck in the 19th century, involving the inheritance of acquired characteristics as an alternative to Darwin's mechanism of natural selection in evolution; the concept, however, was only part of a much broader evolutionary theory developed by Lamarck
Little Ice Age	an unusually cool period, but not as cold as the glacial ice ages in the Earth's past, that lasted from about 1500 to the beginning of modern global warming around 1850
low-fat diet	a diet that limits total fat intake to between 20% and 35% of calories, with no more than 10% from saturated fat, according to recent versions of *Dietary Guidelines*; the so-called prudent low-fat diet places no limit on total fat but restricts saturated fat as well as dietary cholesterol
MMR	measles, mumps, and rubella
MRFIT	Multiple Risk Factor Intervention Trial
NAS	U.S. National Academy of Sciences
NASA	U.S. National Aeronautics and Space Administration
natural selection	the mechanism that governs survival in Darwin's theory

	of evolution, involving nature's selection of the genetic variations best suited to reproduction of the species
NHI	U.S. National Heart Institute
NHLBI	U.S. National Heart, Lung, and Blood Institute
NIH	U.S. National Institutes of Health
NOAA	U.S. National Oceanic and Atmospheric Administration
NSF	U.S. National Science Foundation
Paris Agreement	a worldwide agreement that obligates its signatories to declare voluntary contributions toward reducing emissions of CO_2 and other greenhouse gases, with the stated objective of limiting the temperature increase purportedly resulting from such emissions; the agreement became effective in 2016
plate tectonics	a theory explaining the dynamics of the earth's crust, in terms of motion of 15 to 20 thick, rigid overlapping plates that glide very slowly over the underlying mantle
radioactive	spontaneous decay of an unstable atomic nucleus, in which the nucleus emits one of several possible types of radiation; radioactive decay of certain chemical elements such as uranium can be used for dating ancient rocks and fossils
transgene	a foreign gene transferred to the genome of a host organism, such as a plant, in genetic engineering
UNFCCC	UN Framework Convention on Climate Change
USDA	U.S. Department of Agriculture
VAERS	U.S. Vaccine Adverse Event Reporting System
young-Earth creationism	
	a form of creationism whose proponents believe that the natural world was created by God only 6,000 to 10,000 years ago, and that the earth's geological features were formed by the great flood described in the Bible
WHI	Women's Health Initiative
WMO	World Meteorological Organization

BIBLIOGRAPHY

Alexander, Ralph B. 2012. *Global Warming False Alarm: The Bad Science behind the United Nations' Assertion that Man-Made CO₂ Causes Global Warming*. Royal Oak, MI: Canterbury Publishing.

Barlow, Nora. 1958. *The Autobiography of Charles Darwin 1809–1882*. London: Collins.

Biss, Eula. 2014. *On Immunity: An Inoculation*. Minneapolis, MN: Graywolf Press.

Blackwell, Richard J. 2008. *Behind the Scenes at Galileo's Trial*. Notre Dame: University of Notre Dame Press.

Bowler, Peter J. 1996. *Charles Darwin, The Man and His Influence*. Cambridge: Cambridge University Press.

Bray, Dennis and Hans von Storch. 2016. *The Bray and von Storch 5th International Survey of Climate Scientists 2015/2016*, HZG Report 2016-2. Geesthacht, Germany: Helmholtz-Zentrum Geesthacht.

Cohen, H. Floris. 1994. *The Scientific Revolution: A Historiographical Inquiry*. Chicago: University of Chicago Press.

Copi, Irving M., Carl Cohen and Kenneth McMahon. 2015. *Introduction to Logic*. Boston: Pearson.

Coyne, Jerry A. 2010. *Why Evolution Is True*. London: Penguin Books.

Darwin, Charles. 1845. *Journal of Researches into the Natural History and Geology of the Countries Visited During the Voyage of H.M.S. Beagle Round the World, Under the Command of Capt. FitzRoy, R.N.* London: John Murray, Albemarle Street.

_____, 1859. *On the Origin of Species by Means of Natural Selection, or the Preservation of Favoured Races in the Struggle for Life*. London: John Murray, Albemarle Street.

_____, 1871. *The Descent of Man, and Selection in Relation to Sex*. London: John Murray, Albemarle Street.

Dawkins, Richard. 1996. *The Blind Watchmaker: Why the Evidence of Evolution Reveals a Universe without Design*. New York: W. W. Norton & Company.

de Solla Price, Derek J. 1963. *Little Science, Big Science*. New York: Columbia University Press.

DiMento, Joseph F. C. and Pamela Doughman (eds.). 2007. *Climate Change: What It Means for Us, Our Children, and Our Grandchildren*. Cambridge, MA: The MIT Press.

Druker, Steven M. 2015. *Altered Genes, Twisted Truth: How the Venture to Genetically Engineer Our Food Has Subverted Science, Corrupted Government, and Systematically Deceived the Public*. Salt Lake City: Clear River Press.

Durbach, Nadja. 2004. *Bodily Matters: The Anti-Vaccination Movement in England, 1853-1907*. Durham, NC: Duke University Press.

Farrington, Benjamin. *Greek Science*. 1981. Chester Springs, PA: Dufour Editions.

Fedoroff, Nina V. and Nancy Marie Brown. 2004. *Mendel in the Kitchen: A Scientist's View of Genetically Modified Foods*. Washington, DC: Joseph Henry Press.

Feynman, Richard P. 1999. *The Pleasure of Finding Things Out: The Best Short Works of Richard P. Feynman*. Cambridge, MA: Perseus Books.

Frankel, Henry R. 2012. *The Continental Drift Controversy, Vol. I: Wegener and the Early Debate*. Cambridge: Cambridge University Press.

_____, 2012. *The Continental Drift Controversy, Vol. II: Paleomagnetism and Confirmation of Drift*. Cambridge: Cambridge University Press.

_____, 2012. *The Continental Drift Controversy, Vol. III: Introduction of Seafloor Spreading*. Cambridge: Cambridge University Press.

_____, 2012. *The Continental Drift Controversy, Vol. IV: Evolution into Plate Tectonics*. Cambridge: Cambridge University Press.

Fuller, Watson (ed.). 1971. *The Social Impact of Modern Biology*. London: Routledge & Kegan Paul.

Garrison, Fielding H. 1966. *An Introduction to the History of Medicine*. Philadelphia: W.B. Saunders Company.

Gelernter, David. 2018 (to be published). *Subjectivism: The Mind from Inside*. New York: W. W. Norton & Company.

Gould, Stephen Jay. 2002. *Rocks of Ages: Science and Religion in the Fullness of Life.* New York: Ballantine Books.

Greene, Mott T. 2015. *Alfred Wegener: Science, Exploration, and the Theory of Continental Drift.* Baltimore: Johns Hopkins University Press.

Harris, Richard. 2017. *Rigor Mortis: How Sloppy Science Creates Worthless Cures, Crushes Hope, and Wastes Billions.* New York: Basic Books.

Henig, Robin Marantz. 2000. *The Monk in the Garden: The Lost and Found Genius of Gregor Mendel, the Father of Genetics.* New York: Houghton Mifflin Company.

Henderson, D. A. 2009. *Smallpox: The Death of a Disease — The Inside Story of Eradicating a Worldwide Killer.* Amherst, NY: Prometheus Books.

Huxley, Leonard. 1900. *Life and Letters of Thomas Henry Huxley, Vol. I.* New York: D. Appleton and Company.

IPCC. 1996. *Climate Change 1995: The Science of Climate Change. Contribution of Working Group I to the Second Assessment Report of the Intergovernmental Panel on Climate Change*, J. T. Houghton, L. G. Meira Filho, B. A. Callander et al. (eds.). Cambridge: Cambridge University Press.

IPCC. 1996. *Climate Change 1995: Impacts, Adaptations and Mitigation of Climate Change: Scientific-Technical Analyses. Contribution of Working Group II to the Second Assessment Report of the Intergovernmental Panel on Climate Change*, R. T. Watson, M. C. Zinyowera, R. H. Moss and D. J. Dokken (eds.). Cambridge: Cambridge University Press.

IPCC. 2001. *Climate Change 2001: The Scientific Basis. Contribution of Working Group I to the Third Assessment Report of the Intergovernmental Panel on Climate Change*, J. T. Houghton, Y. Ding, D. J. Griggs et al. (eds.). Cambridge: Cambridge University Press.

IPCC. 2007. *Climate Change 2007: Mitigation. Contribution of Working Group III to the Fourth Assessment Report of the Intergovernmental Panel on Climate Change*, B. Metz, O. R. Davidson, P. R. Bosch et al. (eds.). Cambridge: Cambridge University Press.

IPCC. 2013. *Climate Change 2013: The Physical Science Basis. Working Group I Contribution to the Fifth Assessment Report of the Intergovernmental Panel on Climate Change*, T. F. Stocker, D. Qin, G.-K. Plattner et al. (eds.). Cambridge: Cambridge University Press.

Keys, Ancel (ed.). 1970. *Coronary Heart Disease in Seven Countries.* New York: The American Heart Association, Monograph No. 29.

Krips, Henry, J. E. McGuire and Trevor Melia (eds.). 1995. *Science, Reason, and Rhetoric.* Pittsburgh: University of Pittsburgh Press.

Kuhn, Thomas S. 2012. *The Structure of Scientific Revolutions.* Chicago: University of Chicago Press.

Lindberg, David C. 2008. *The Beginnings of Western Science: The European Scientific Tradition in Philosophical, Religious, and Institutional Context, Prehistory to A.D. 1450.* Chicago: University of Chicago Press.

Mangelsdorf, Paul C. 1975. "Donald Forsha Jones: April 16, 1890—June 19, 1963." In U.S. National Academy of Sciences, *Biographical Memoirs, Volume XLVI,* Washington, DC: National Academies Press.

McClellan, James E. III and Harold Dorn. 2006. *Science and Technology in World History: An Introduction.* Baltimore: Johns Hopkins University Press.

Michaels, Patrick J. and Paul C. Knappenberger. 2016. *Lukewarming: The New Climate Science that Changes Everything.* Washington, DC: Cato Institute.

Miller, Neil Z. 2016. *Miller's Review of Critical Vaccine Studies: 400 Important Scientific Papers Summarized for Parents and Researchers.* Santa Fe, NM: New Atlantean Press.

Monod, Paul Kléber. 2013. *Solomon's Secret Arts: The Occult in the Age of Enlightenment.* New Haven: Yale University Press.

NIPCC. 2013. *Climate Change Reconsidered II: Physical Science,* C. D. Idso, R. M. Carter and S. F. Singer (eds.). Chicago: The Heartland Institute.

NIPCC. 2014. *Climate Change Reconsidered II: Biological Impacts,* C. D. Idso, S. B. Idso, R. M. Carter and S. F. Singer (eds.). Chicago: The Heartland Institute.

Numbers, Ronald L. 1992. *The Creationists.* New York: Alfred A. Knopf.

Offit, Paul A. 2015. *Deadly Choices: How the Anti-Vaccine Movement Threatens Us All.* New York: Basic Books.

Offit, Paul A. and Charlotte A. Moser. 2011. *Vaccines and Your Child: Separating Fact from Fiction.* New York: Columbia University Press.

Opie, Iona and Peter (eds.). 1997. *The Oxford Dictionary of Nursery Rhymes.* Oxford: Oxford University Press.

Paley, William. 1802. In Frederick Ferré, ed., *Natural Theology: Selections,* Indianapolis: The Bobbs-Merrill Company, 1963.

Pennock, Robert T. (ed.). 2001. *Intelligent Design Creationism and Its Critics: Philosophical, Theological, and Scientific Perspectives.* Cambridge, MA: The MIT Press.

Petto, Andrew J. and Laurie R. Godfrey (eds.). 2007. *Scientists Confront Intelligent Design and Creationism.* New York: W. W. Norton & Company.

Popper, Karl. 2014. *The Logic of Scientific Discovery.* Eastford, CT: Martino Fine Books.

Pringle, Peter. 2005. *Food, Inc.: Mendel to Monsanto — The Promises and Perils of the Biotech Harvest.* New York: Simon & Schuster.

Prothero, Donald R. 2007. *Evolution: What the Fossils Say and Why It Matters.* New York: Columbia University Press.

Robinson, Daniel N., Gladys Sweeney and Richard Gill (eds.). 2006. *Human Nature in Its Wholeness: A Roman Catholic Perspective.* Washington, DC: The Catholic University of America Press.

Sagan, Carl. 1997. *The Demon-Haunted World: Science as a Candle in the Dark.* New York, Ballantine Books.

Scott, Eugenie C. 2009. *Evolution vs. Creationism: An Introduction.* Berkeley: University of California Press.

Shermer, Michael. 2015. *The Moral Arc: How Science Makes Us Better People.* New York: Henry Holt and Company.

Smallman-Raynor, Matthew and Andrew Cliff. 2012. *Atlas of Epidemic Britain: A Twentieth Century Picture.* Oxford: Oxford University Press.

Teicholz, Nina. 2014. *The Big Fat Surprise: Why Butter, Meat and Cheese Belong in a Healthy Diet.* New York: Simon & Schuster.

The Royal Society and U.S. National Academy of Sciences. 2014. *Climate Change: Evidence & Causes.* Washington, DC: National Academies Press.

UK National Advisory Committee on Nutrition Education. 1983. *A Discussion Paper on Proposals for Nutritional Guidelines for Health Education in Britain.* London: Health Education Council.

UN Food and Agriculture Organization. 2010. *Fats and Fatty Acids in Human Nutrition: Report of an Expert Consultation, 10-14 November 2008, Geneva,* Food and Nutrition Paper 91. Rome: Food and Agriculture Organization of the United Nations.

U.S. Department of Agriculture. 1980. *Nutrition and Your Health: Dietary Guidelines for Americans.* Washington, DC: U.S. Government Printing Office.

U.S. Institute of Medicine, Board on Population Health and Public Health Practice. 2012. *Adverse Effects of Vaccines: Evidence and Causality;* Kathleen Stratton, Andrew Ford, Erin Rusch and Ellen Wright Clayton (eds.). Washington, DC: National Academies Press.

U.S. Institute of Medicine and National Research Council. 2004. *Safety of Genetically Engineered Foods: Approaches to Assessing Unintended Health Effects.* Washington, DC: National Academies Press.

U.S. National Academy of Sciences. 1975. *Biographical Memoirs, Volume XLVI.* Washington, DC: National Academies Press.

U.S. National Academies of Sciences, Engineering, and Medicine. 2015. *Trust and Confidence at the Interfaces of the Life Sciences and Society: Does the Public Trust Science? A Workshop Summary*. Washington, DC: National Academies Press.

U.S. National Academies of Sciences, Engineering, and Medicine. 2016. *Genetically Engineered Crops: Experiences and Prospects*. Washington, DC: National Academies Press.

U.S. National Research Council. 1979. *Carbon Dioxide and Climate: A Scientific Assessment*, Report of an Ad Hoc Study Group on Carbon Dioxide and Climate. Washington, DC: National Academies Press.

U.S. National Research Council. 1980. *Toward Healthful Diets*. Washington, DC: National Academies Press.

U.S. National Research Council. 2006. *Surface Temperature Reconstructions for the Last 2,000 Years*. Washington, DC: National Academies Press.

U.S. National Science Board. 2016. *Science and Engineering Indicators 2016*. Arlington, VA: National Science Foundation.

U.S. Senate Committee on Environment and Public Works. 1986. *Ozone Depletion, The Greenhouse Effect, and Climate Change*. Washington, DC: U.S. Government Printing Office.

U.S. Senate Select Committee on Nutrition and Human Needs. 1977. *Dietary Goals for the United States*. Washington, DC: U.S. Government Printing Office, 2nd Edition, December 1977.

van der Gracht, W. van Waterschoot (ed.). 1928. *Theory of Continental Drift: A Symposium on the Origin and Movement of Land Masses Both Inter-continental and Intra-Continental, as Proposed by Alfred Wegener*. Tulsa: American Association of Petroleum Geologists.

Zenko, Micah. 2015. *Red Team: How to Succeed by Thinking Like the Enemy*. New York: Basic Books.

Aristarchus 8, 13
Aristotle 6, 7, 8, 10, 11, 12, 13, 14, 15, 16, 17, 19, 21, 24, 26, 116, 221
Arrhenius, Svante 108, 121
artificial intelligence 161
ASCN 91, 105
Asia 32, 116, 185
AstraZeneca 187, 189
astrology 11, 19, 26
astronomy 6, 7, 10, 11, 13, 14, 15, 18, 26, 30, 44
atherosclerosis 80, 81, 89, 100, 101
Atlantic Multidecadal Oscillation 123, 135
Atlantic Ocean 31, 47, 60
atmosphere 30, 73, 108, 112, 121, 125, 130
  stratosphere and 128, 129
  troposphere and 122, 128, 129, 130
atolls 40, 51, 52
Australia 32, 38, 39, 45, 46, 57, 58, 69, 70, 137, 165
Austria 187
authority 18, 44
  of the church 13, 15, 25, 44, 59, 117
  of the scientific establishment 25, 29, 44, 59, 79, 102, 104, 116, 117, 180, 185, 196, 211
autism 151, 153, 157, 158, 159, 160, 161, 164, 169, 184, 203, 213

B

babies 117, 146, 161, 163, 190
Bacon, Sir Francis 20, 24
bacterial flagellum 67, 74
barley 172
Bayer 202
Beagle voyage 50, 51, 52, 54, 58, 69
Behe, Michael 66, 67, 73, 74, 75
beta-carotene 185, 186, 187
Beyer, Peter 185, 186, 187, 191
bias 22, 46, 82, 103, 104, 105, 114, 127, 192, 206, 209, 212. *See* also confirmation bias
Bible 12, 15, 16, 19, 29, 59, 60, 61, 64,

65, 66, 73, 200
Biblical flood 59, 63, 64, 70, 71, 72
biogeography 57, 58, 70, 125
biological evolution
  descent with modification in 50, 52, 58
biology 7, 18, 21, 28, 43, 47, 50, 51, 54, 60, 62, 63, 67, 68, 70, 74, 75, 125, 200, 202, 219
  textbooks 60, 62, 63, 66, 71
birds 57, 58, 63, 64
  finches 58
Blackburn, Henry 104, 106
Black Death 13, 143
Blackett, Patrick 36, 37
blood clotting cascade 74
Boeing study 95
Brahe, Tycho 14, 26
Brazil 191
Brazil nuts 178, 192
British Association for the Advancement of Science 55
British Medical Journal 103, 104, 204, 214
Bronowski, Jacob 209
Browner, Warren 100
Bryan, William Jennings 60, 61
Bt corn 181, 182, 183, 184, 192
Bt insecticides 181, 182, 183
  event 176 hybrid 182
Burbank, Luther 172
bureaucracy 105

C

calculus 19, 44
CaMV 35S 176, 177, 183, 184, 196, 203. *See* also genetic engineering
Canada 69, 113, 151, 165, 168, 177, 181, 191, 194
cancer 91, 94, 95, 96, 97, 105, 157, 176, 177, 194, 196, 202, 203, 208, 218
  breast cancer 94, 95, 97, 203
  leukemia 167
  lung cancer 86, 105, 136
  prostate cancer 97

embryology 58
emissions
aerosol 108, 121, 124
chlorofluorocarbon 129
CO_2 108, 112, 113, 114, 116, 117, 119,
126, 132, 134, 135, 136, 137, 141,
201, 207
encephalitis 150, 152, 165
engineering 7, 23, 26
England 18, 19, 21, 43, 52, 59, 144,
145, 146, 147, 149, 152, 158, 191
EPA 126, 188, 195
epidemics 147, 149
measles 155, 156, 165, 167, 168
smallpox 144, 147
whooping cough 150
epidemiological studies 82, 84, 85,
86, 87, 88, 97, 101, 106, 125, 136,
158, 161, 194, 202, 207, 212
confounding factors in 85, 97
Euclid 6, 12
Europe 10, 11, 12, 13, 19, 26, 27, 30, 31,
32, 38, 46, 59, 81, 83, 115, 135, 144,
145, 187, 188, 189, 191, 193, 194,
201, 209, 218
European Chemicals Agency 188
European corn borer 181
evidence
anecdotal 209, 217
empirical 5, 7, 21, 25, 43, 52, 56,
58, 60, 69, 70, 107, 117, 120,
123, 135, 136, 139, 141, 142, 199,
200, 201, 206, 207, 220, 221
medical 159, 162, 164, 166
mistreatment of 62, 65, 66, 68, 70,
75, 76, 77, 79, 88, 104, 107, 115,
116, 136, 199, 200, 210, 216
scientific observations as 14, 15,
24, 31, 36, 37, 38, 40, 41, 42,
54, 70, 73, 105, 106, 125, 131,
132, 134, 160, 161, 164, 168, 176,
201, 215
suppression of 49, 99, 140, 210,
214
evolution. *See* theories: theory of
evolution

extreme weather 136
eye, human 50, 75

F

faith 18, 25, 73, 74, 76, 77, 139, 200,
201, 205, 215, 220
falsifiability 24, 25, 76
faunal succession 63, 71
FDA 159, 177, 178, 180, 188, 190, 194,
195
generally recognized as safe
(GRAS) criterion of 194
fear 5, 138, 140, 143, 145, 146, 164, 166,
169, 171, 176, 177, 188, 192, 193,
196, 201, 207, 208
Feynman, Richard 23, 25, 217
Finland 82, 83, 153
fish 58, 63, 64, 71, 91, 159, 177, 178
Fisher, Barbara Loe 153
flood geology 63, 64, 65, 70, 71, 72,
73, 75, 76, 200
floods 135
flying squirrel 58
formaldehyde 154, 163
fossils
as evidence for continental drift
31, 33, 35, 36, 43, 47, 199
as evidence for evolution 50, 52,
56, 57, 58, 60, 69, 70, 71, 72,
76, 200
in geologic column 63, 64, 65, 70
Framingham Heart Study 85, 87, 89,
99, 100, 200
France 18, 81, 83, 187
Frankenfood 171, 178, 186, 192, 197
funding 7, 22, 26, 27, 28, 79, 83, 89,
100, 102, 105, 106, 138, 139, 187,
209, 210, 211, 218, 219
by city states 26
by court patronage 8, 26, 27
by government 28
by industry 27
by institutions 27
by nation states 26
for applied research 27

USDA 90, 91, 92, 94, 105, 177, 182,
195, 197, 200, 203, 206
U.S. state laws
against teaching of evolution 60,
61, 62, 65, 206
requiring equal time for creation
science and evolution 65, 68,
70, 206

V

vaccination 5, 192, 201, 202, 206, 213,
214, 215
arm to arm 145, 146
lawsuits against 150, 152, 156, 157,
158, 169
mandatory 145, 146, 147, 155, 156,
167, 168, 196, 206
medical exemptions from 157, 169
objections to, by Christian Scientists 156
philosophical exemptions from
143, 157, 167, 168
rate of 146, 149, 166, 167, 168
religious exemptions from 143,
156, 157, 167, 168
risk of adverse reaction to 143,
148, 150, 151, 152, 154, 155, 161,
163, 164, 165, 166, 167
risk of disease without 148, 149,
152, 165, 166, 167
side effects of 143, 145, 146, 148,
149, 150, 152, 153, 154, 157,
162, 163, 164, 165, 166, 167,
192, 201, 203
vaccines
aluminum in 164
DTP 149, 150, 151, 152, 153, 155,
157, 159, 169, 203, 213
Hib 153, 157
influenza 159, 163
killed virus 154, 155, 163
live-virus 144, 154, 155, 165
MMR 149, 152, 153, 154, 155, 156,
157, 158, 159, 160, 161, 168, 169,
184, 203, 213
polio 149, 154, 155, 156, 165

rabies 148
rotavirus 154, 155, 162, 165
safety issues with 155, 162, 164,
165, 166, 215
smallpox 145, 148
whooping cough 203
VAERS 162, 163
Van Montagu, Marc 193
vestigial structures 58, 59
Villach conferences 110, 111, 137
Vine, Frederick 41
vitamin A deficiency 185, 186, 187
volcanic rocks 37
volcanoes 37, 39, 40, 63, 72, 120, 121,
129

W

Wakefield, Andrew 153, 157, 158,
159, 160, 161, 164, 169, 184, 203,
213
Wales 146, 147, 149, 190, 191, 193
Wallace, Alfred Russell 54
Washington, General George 144
watchmaker argument 50, 51, 53, 66
water vapor 110, 121, 122, 123, 125
weather extremes 134, 135
Wegener, Alfred 29, 30, 31, 32, 33,
34, 35, 36, 38, 39, 41, 42, 43, 44,
45, 46, 47, 48, 57, 59, 116, 140,
169, 196, 199, 200, 213, 214, 215
wheat 172, 188, 190, 191, 219
whooping cough 143, 148, 149, 151,
152, 164, 168
Wilberforce, Bishop Samuel 55
Willett, Walter 103, 104
Wilson, John 149, 150, 151, 153, 203
Wirth, Senator Timothy 110
Womens Health Initiative 94, 95,
96, 97, 98, 200

Y

Yerushalmy, Jacob 81, 82, 83, 88, 101